'This book is a kind of miracle, a thrilling compendium of plays that speak to the enormous environmental crisis of our time. Freestone and O'Hare have exquisite taste and brilliant analysis, illuminating plays I've never heard of, as well as plays I thought I knew. *100 Plays to Save the World* should be required reading for everyone who believes in the power of theatre to move the world; I will certainly never plan a season again without referring to it.' *Oskar Eustis, Artistic Director, The Public Theater, New York*

'These are hugely important plays that address the biggest and scariest thing mankind has ever faced. Words that everyone should be reading and seeing on our stages. A vital resource for us all.' *Morgan Lloyd Malcolm, playwright* (*Emilia, Mum*)

'This remarkable new volume accomplishes three essential tasks in the ongoing attempt to amplify the relationship between theatre and environment: it highlights the playwright's agency, as much as the audience's; it builds a bridge connecting past, present and future; it takes an internationalist approach. In so doing, it shows us that environment and climate are collective causes and crises that require unity, engagement and intervention, from both sides of the stage, and across borders.' *Vicky Angelaki, Professor in English Literature, Mid Sweden University*

'An essential, eclectic, and expertly contextualised trove of plays reflecting our delicate, changing relationship with the natural world.' *Rupert Goold, Artistic Director, Almeida Theatre, London*

'This book is dynamite. Through lively play analysis and accessible environmental know-how, it will galvanise theatre-makers to step up and artists to be heard. Theatre must play its part in the climate fight and this book shows us how.' *Kwame Kwei-Armah, Artistic Director, Young Vic Theatre, London*

'Whatever the opposite of boring and worthy is, this book is it.' *Nicola Walker, actor* (*The Curious Incident of the Dog in the Night-Time*, *A View from the Bridge*, *Last Tango in Halifax*, *Unforgotten*)

Elizabeth Freestone
and Jeanie O'Hare

100 PLAYS TO SAVE THE WORLD

THEATRE COMMUNICATIONS GROUP
NEW YORK
2023

100 Plays to Save the World is published by Theatre Communications Group, Inc., 520 Eighth Avenue, 20th Floor, Suite 2000, New York, NY 10018-4156

This volume is published in arrangement with Nick Hern Books Limited, The Glasshouse, 49a Goldhawk Road, London W12 8QP

The publication of *100 Plays to Save the World* by Elizabeth Freestone and Jeanie O'Hare, through TCG Books, is made possible with support by Mellon Foundation.

TCG books are exclusively distributed to the book trade by Consortium Book Sales and Distribution.

ISBN 978-1-63670-144-8 (paperback) / ISBN 978-1-63670-214-8 (ebook)

A catalog record for this book is available from the Library of Congress.

Cover design by Mark Melnick
Cover photo by Jonathan Pozniak/Gallery Stock

First TCG Edition, October 2023

Contents

Foreword

The arts have an essential and urgent role to play in shaping our world toward a regenerative future. Ask any climate scientist and they'll tell you: we're on the precipice of interlocking crises of immense scale, and what we have is a narrative problem. Our societal narratives speak of doom and gloom, economic recession and disaster capitalism as though they're inevitable. We hear over and over about all we stand to lose in the coming decade. But what if we could change that narrative... what if we could *re-story* the future?

What a gift to the field that this collection has been organized to lead us from destruction to hope.

This constellation of essays about climate-focused plays provides an incredibly rich resource to draw upon and a fantastic starting place for those who may not know where to begin when engaging with the immensity of the climate crisis. Spanning over a century, from the Industrial Age to new works that have yet to be published, the stories included here speak to every theatre-maker and audience member who longs for clean air, fresh water, and a future for their families.

The deep knowledge that lives in these plays is firmly rooted in generations of work that have been led by Black and Indigenous peoples. The climate crisis is not a singular issue, as many of these plays explore; it is part of and results from interconnecting systems of oppression, requiring action on all our parts to dismantle white supremacy, anti-Blackness, settler colonialism, racism, ableism, and much more. As you experience these works, we invite you to sit while unsettled, reflect while challenged, and engage in the many calls to action written across these pages.

At Groundwater Arts, through our work organizing for a Green New Theatre, we know that it's not just about the stories we tell, but how we tell them. The way we currently create theatre is not sustainable. The field is depleting resources faster than the earth can replenish them, and we are generating more than our fair share of waste. Producing or being impacted by a powerful piece of theatre is not enough. We must also, as this book invites us, move from reading plays to taking action in our communities; move from extractive systems to regenerative ones; move from destruction to hope.

Who better than artists and storytellers to re-shape the narrative? Theaters and theatre-makers must contribute to shifting societal stories and creating the futures we know we deserve.

Annalisa Dias and Tara Moses
Co-Founders of Groundwater Arts

Annalisa Dias is a Goan-American transdisciplinary artist, community organizer, and award-winning theatre-maker. She is Director of Artistic Partnerships & Innovation at Baltimore Center Stage and a Co-Founder of Groundwater Arts. www.annalisadias.weebly.com

Tara Moses is a citizen of Seminole Nation of Oklahoma, Mvskoke, a director, multi-award-winning playwright, consultant, and Co-Founder of Groundwater Arts. www.taramoses.com

Groundwater Arts is a US-based artist collaborative committed to re-envisioning the arts field through a climate justice lens. Our decolonial approach is activated through the principles of a Green New Theatre (GNT), a movement-building document penned by Groundwater Arts in partnership with countless arts makers throughout the country. Our mission is to shape, steward, and seed a just* future through creative practice, consultation, and community building. www.groundwaterarts.com

* climate justice = racial justice = economic justice = a decolonized future

This book is for our parents and grandparents,
who taught us to care for the world,
and for the younger members of our family,
who taught us why we must keep caring.

Introduction

People often ask: where is the great climate-change play? The answer is it's here, it has already been written, and quite possibly it was staring you in the face. Writers have for years been wrestling with the challenges the world now faces, but clarion calls from the past by visionary playwrights are only now being listened to. Extinction, extreme weather, resource shortages, failing political leadership, truth, denial – these things already exist in the playwriting culture. We just need a sharp new ear to tune into their resonances. In addition, new plays are being written every day dealing head-on with these topics. This book attempts to encourage new readings of those classic plays and to propel great new writing into sharper focus.

This book is unapologetically activist. It is a call to arms. The aim is to make programming great environmental plays easier, watching great environmental plays more accessible and having life-changing conversations much more likely. This book is not intended to be prescriptive. Instead, we offer it as an empowering and enabling guide. We hope it is of use to theatre-makers in rural and urban contexts, in venues and on tour, student, amateur and professional. All of us are in the same fight. We are imagining progressive new productions of these plays that inspire new thinking. We hope you will not be limited by place or time or setting. We encourage you to make dynamic, theatrical choices that offer audiences fun, provocation and hope, and most importantly, the possibility of change.

We – artists, thinkers, creators – have a responsibility to communicate the truth of this emergency. The future we currently face is as uncertain as it is daunting. The world is shape-shifting and our culture must too. We believe by wrestling with some of the issues raised in these plays we can help tell new stories about the way we might want to live. There are plays for all tastes here, all points of view, all approaches to this cultural moment. This selection of plays, seen in context with each other, stimulates a conversation about how many ways we can invent our future.

The Anthropocene is the name given to the geological age we are in now. Named after the Greek '*anthropos*', meaning 'man', it was chosen to emphasise the truth that humankind has now left a geological footprint on this planet: radioactive isotopes are found in glacial ice; the high levels of CO_2 in the atmosphere are detectable in tree rings and limestone; our plastic waste is forming a new sedimentary layer. But still large swathes of the population opt out of believing in these facts. Why? We have to consider that the stories we tell, the way in which we tell them, and on which stages they are told, might be part of the problem. We urge theatre-makers and programmers to become part of the Theatre of the Anthropocene, telling stories that anticipate our future, acknowledge our past and make our present liveable.

Fighting the climate crisis is a global endeavour. We read plays from English-speaking theatre cultures, as well as many plays in translation. We called theatre companies, spoke to agents, contacted writers directly, sought recommendations and searched online. This book includes writers from many countries and many different backgrounds. There are voices and places under-represented – and we urge translators and commissioners to enable more work from the Global South to be heard. Excitingly, new work about the environment is being produced all the time. It may be strange to say, but we are hopeful this book will date rapidly and that soon thousands of new plays about our new world reality will dominate our global stages. This book is simply a snapshot of where we are now and shows us where we need to go. Our criteria was that each play needed to have been/be about to be professionally produced and/or be available in published/accessible form. We sought a wide range of narrative styles, small casts and large, ensemble and leading roles, domestic and political, intergenerational, animals and human characters. The idea that all climate plays are either scientist plays, dystopias, or had to have a polar bear in them was blown clear out of the water very early on.

Some of the plays included here were written long before such a thing as a climate crisis was known about. Some were written more recently but without an explicitly stated intention that the play addresses environmental issues. In both instances we have cheerfully taken the view that works of art out there in the world are to be interpreted and interrogated as others see fit. We don't pretend to put forward definitive readings of these plays, simply to allow aspects of them that we believe could speak to this moment to shine through. Relationships to nature, geopolitical issues, social consequences of environmental impacts; all of these, to us, help tell the story of the most pressing issue facing us today.

The plays chosen from further back in time – Aristophanes, Chekhov, Brecht – seem eerily prescient when read through environmental eyes, both predicting and speaking directly to this moment. Their relevance is a useful reminder that staging environmental stories is not just the responsibility of playwrights. Theatre-makers of every discipline – casting, design, acting, directing, stage management – must reimagine and reinterpret these plays through the prism of the present. The contemporary plays – by writers such as Erika Dickerson-Despenza, Chris Bush and Lucy Kirkwood – concentrate their fire on a diversity of targets, visionary in their writing and unflinching in their gaze. The climate crisis is not one problem. Turning down the global thermostat won't solve habitat destruction or reconnect people to the natural world. Collectively, we hope the choice of plays creates space for dynamic discussion of the multifaceted, interconnected, complex collision of environmental challenges we are now facing.

The compiling of this collection has reminded us that we need to acknowledge that the nature of our international theatre reveals our collective thinking, and that maybe our collective thinking is sleepily behind the curve. The wilder plays are born from the imaginations of writers whose neighbourhoods are burning and whose homes are flooding. The further you are from the daily lived reality of the climate crisis, the quieter and more formally conservative the plays. This collection has revealed how the world is reshaping itself violently in the physical realm and how that is impacting on the reshaping of stories we need to tell, not just for now but for generations to come. This climate emergency will, in many ways, be the subject of all of our art for the foreseeable future, just as it ought to be the dominant discourse in our political, economic and social spheres.

Writers won't just write plays about these issues for a short while, after a fashion, believing the crisis will then be over. This is our new reality. The shifts we make societally in the next decade will be with us forever, otherwise the undeniable truth is that the concept of forever will itself no longer exist.

We have learned from assembling this collection that the impact of the climate emergency is altering the way that plays are written and for whom they are written. The movement of peoples has an impact on our stories, and the rise in the pitch of the voices that need to be heard has an impact on our listening.

This collection has also raised questions about how epic the plays need to be, how global in outlook, how formally elastic and inventive. We can no longer navel-gaze and clink our gins. We need to capture a reality that we have never experienced before. We need to unleash the power of a total theatre, an era of playwriting that embraces epic stories, and values playwrights' intelligent, focused urgency and understanding. We need to exercise and stretch our thinking, widen our eyes, strengthen our neck muscles for the sustained looking up we now need to do. Theatre must imagine the future, and help us reach towards the bold, humane, quick thinking we are going to need.

About This Book

We have focused on playwrights. We know there is much brilliant devised, unscripted, immersive, interactive, site-responsive work out there, and we salute and applaud all of those makers. The intention here is simply for producers and companies to find scripts they can read and plays they can consider programming.

The plays included have all been written in or translated into English. They are available either in published form, online, or through direct contact with writers. Libraries, bookshops and search engines will lead the way.

We limited ourselves to one play per writer. There were multiple plays by Caryl Churchill and Wole Soyinka, for example, that we could have chosen. But given the wealth of material on offer, we decided to feature as many playwrights as possible, introduce as many writers to readers as possible, support the broadest range of artists that we could, and trust the reader to go on a subsequent journey of their own discovery.

Apart from the odd exception, we have focused on full-length plays. There are brilliant initiatives that ask writers from around the world to contribute short plays for instant debate and easy dissemination. More information about these can be found at the end of the book.

We read well over three hundred plays and could've read many more. This is not a definitive collection. We chose strong plays that we feel speak to this moment in theatrically exciting ways. We wanted an overall balance of content, genre and cast size. We wanted to represent as diverse a group of ideas and writers as possible. We know there are lots of plays we left out and we embrace the debate. Get in touch, tell us, social media the hell out of it. Let's collectively put together another book of another hundred.

We spent a lot of time wondering how to arrange the plays and finally settled on twelve chapters, each representing a particular aspect of the climate crisis. Each chapter carries a short introduction by way of establishing some context for the situation we find ourselves in. Although each play is situated within one of these categories, the chapters are offered as loose collectives. Don't let that stop you from focusing on other aspects of the play in your productions. Some plays could've been put in two or more categories. Others are category-defying by nature.

At the back of the book are indexes of the plays by title, playwright's name and cast size to give you alternative ways to search for plays you might be interested in. There are also resources for further reading and some information on how this book was printed.

Otherwise, we recommend flicking through, landing on a play, planning a production, and saving the world.

Thanks

Kirsten Adam, Frédérique Aït-Touati, Martin Banham, Matt Connell, Pierre Daubigny, Hart Fargo, Sonia Fernandez, Michael Finkle, James Gibbs, Judith Greenwood, Maria Manuela Goyanes, Kerry Harvey-Piper, Jennifer Herera, Jim Kleinmann, Charlotte Knight, Julia Kreitman, Eléonore Lamarosse, Chloé Latour, Teya Lazon, Laura Maniura, Julia Mills, Lynn Parker, Jack Phillips Moore, Tori Sampson, Mark Starling, Alexandra Stockley, Olivier Sultan, Jeremy Tiang, Julia Tyrrell, Chloe Vilarrubi, Kennedy Woodard, Derek Zasky and Ayla Zuraw-Friedland all helped us track down plays, source drafts and opened their address books. Particular thanks to Jane Darke and Hamish MacColl who kindly sent us unpublished manuscripts.

Professors Kate Rigby and Sian Sullivan, and Andrea Brogan and Ásta Magnúsdóttir, have been a source of encouragement throughout. Ben and William Freestone gave good advice about the cover art; Jacob Robbins posed useful questions; Rachael Magson supplied tea and cake when it was most needed; and Larkdale Peggy kept us laughing.

Catherine Sheehy, Jenny Pearce and Gemma Bodinetz all read drafts at various stages and offered insightful, rigorous and perfectly timed feedback. Any mistakes in content or tone are ours alone. Rosie Armstrong acted as research aide, proofreader and sounding-board extraordinaire. Her detective skills in tracking down agents/writers/publishers are second to none.

Our thanks must go to Nick Hern and Matt Applewhite for understanding the urgency of the project and supporting both the content and our high-speed writing process. The original UK edition's timely delivery and environmentally conscious printing methodology is testament to their care and attention. Matt's notes and feedback have been invaluable.

Our thanks go to Daze Aghaji for her original foreword to the UK edition. The authors are deeply grateful to Terry Nemeth and everyone at Theatre Communications Group for making this new US edition possible.

Finally, our thanks must go to all the playwrights who sent us their work. Sorry we couldn't include them all. To all of them, and the writers whose work we have included, thank you for grappling with such big and important themes in such thrilling and entertaining ways.

Elizabeth Freestone
and Jeanie O'Hare

PART 1
DESTRUCTION

Destruction

Creativity is supposedly what sets us apart from our fellow animals. We don't have the largest brains or the strongest muscles, we can't fly, or swim great distances underwater. We are hopelessly dependent at birth, taking years to learn how to walk, feed and fend for ourselves. But we can imagine, we can dream, we can draw, write, dance and perform. We can create purely for its own sake, beyond the fulfilment of basic survival needs, and this appears to be an almost exclusively human trait.

For every instinct to create there is also, it seems, the instinct to destroy. No other animal on Earth has caused such irreversible change. No other animal shits where they sleep. We burn and chop, dig and slash, poison and pollute. Then we look away and persuade others to do the same. We are responsible for changing the face of the Earth, but we pretend these actions take place beyond our field of vision, just out of sight. We engage in self-deception about our true character.

The twentieth century has witnessed the most rapacious and accelerated destruction in our planet's history. The six-mile-wide meteor that crashed sixty-six million years ago, killing the dinosaurs, has nothing on us. From the felling of the Amazon rainforest to the bleaching of the Great Barrier Reef, we are ignoring what our stressed-out senses are telling us, that everything we love is dying. All around us, the sea, the landscape, the birds, the fish, the bees, and now Nature's own ability to recalibrate, rebalance and repair is also being destroyed. Her survival mechanisms have reached a tipping point, and a narrative greater than us has now begun. We are no longer in charge, we are now the play-within-the-play. We have no idea how to reverse the narrative of the meta-story. The Earth is sovereign now.

The plays in this chapter offer a broad range of attitudes for meeting this moment. If we are ever to say goodbye to this stubborn part of our collective mindset, we have first to recognise it. From Anton Chekhov's

Uncle Vanya, a piercing look at how the destruction of the environment affects our inner lives, to Alanna Mitchell's thrilling *Sea Sick*, which tells the story of how the oceans have fared on our watch, these plays allow us to re-see, re-set and re-imagine our thinking. Fresh productions of these plays will steer us collectively towards less damaging ways to live.

1 SEA SICK
Alanna Mitchell (2014)

'The truth is, I'm scared. I feel vulnerable, because I'm about to reveal things about myself that I'd really rather you not know. But if not now, when? So I'm gonna do it anyway because I need you to understand what I've found out about the ocean.'

Alanna Mitchell, the writer of this play, is a journalist. She spent three years researching what impact higher levels of carbon dioxide in our atmosphere has on our oceans. What she discovered was unexpected; beautiful and complex ecosystems living under the surface of the water are vital to sustaining the health of the ocean, on which all life on the planet depends. These ecosystems are in a shocking state of decline. Told in the flourishing new form of lecture-demo theatre, she has turned her extraordinary research into a play, using narrative, music, visual aids and metaphor.

We all know how the global thermostat is being notched up year on year by increased carbon-dioxide levels. What is less well known is the effect this has on our planet's oceans. *Sea Sick* looks at what is happening as this colossal life-mass attempts to absorb excess carbon dioxide. The increase in temperature is not the only story. The increase in acidity is turning huge volumes of water into oxygen-depleted pockets, such as the 'dead zone' in the Gulf of Mexico. Without oxygen, there can be no life.

Sea Sick is written with verve and skill. The narrative uncoils like a thriller with a breathless sense of drama. We follow Alanna across continents as she gets her head around why an interest in the health of plankton should be essential for everyone, meets some of the world's poorest communities already being hit by depletion of fish stocks, and witnesses the wonder of nature's unfathomable resilience. Finally, terrifyingly, she descends, in a tiny submersible, to the deepest seabed on the planet. The play's dry humour, unfailing curiosity and refusal to write the human race off just yet makes this a rewarding, uplifting and inspiring story.

2 UNCLE VANYA
Anton Chekhov (1889)

'Forests keep disappearing, rivers dry up, wild life has become extinct, the climate's ruined and the land grows poorer and uglier every day.'

Astrov, the country doctor who utters these words in *Uncle Vanya*, is all too aware of the ecological devastation being wrought by nineteenth-century industrialisation. And yet, like many of Chekhov's characters, he is equally in awe of progress and the social emancipation it promises. The tension between these two ideals snaffles him like flypaper, stuck between inaction and apathy. Astrov's focus closes in on matters of the intellect and the heart, while all around him the sound of the axe, steadily chopping tree after tree, reverberates through his hollowing heart and the emptying landscape. Against the backdrop of an imminent storm, we watch the characters discuss the destruction of the natural environment while they destroy themselves with drink, bemoan unrequited love and mourn unfulfilled dreams. Both the place and its people are on a slippery slope into quiet decline.

Chekhov loved trees. His characters are forever setting off for walks in the woods, noticing the sound of the leaves, or recognising the stupidity of chopping them down. On his own country estate he planted both woods and orchards (yes, including cherry). In *Uncle Vanya*, he presents the full spectrum of human relationships to nature, from characters who respond to aesthetics – 'Oh, isn't that pretty' – to those who want to exploit: 'How much would that bring in?' Chekhov walks a line between the two positions, showing that we may attempt to divorce ourselves from the natural world, but it will always seep deep into our psyches. Underneath everything is a distant alarm bell, ringing somewhere over there, trying to rouse us, attempting to raise into conscious thought the terror of the subconscious: that the impoverishment of the environment is the impoverishment of our souls.

Uncle Vanya is a stunningly beautiful play made more resonant when its central absurd tragedy, of denial and inaction, of fiddling while Rome burns, becomes its wider context.

CONTINUITY 3
Bess Wohl (2019)

'Suddenly we're on a Styrofoam ice shelf in the desert of New Mexico because of tax rebates and the snow is made of plastic and it's a thousand degrees.'

This is a very funny play about what we've done to the world and whether or not we have time to fix it. A film director is trying to get the best take of a scene 'before we lose the light'. She makes five attempts at capturing the shot she needs. She fails each time. The film is set in the Arctic but it's being shot in Mexico for tax-incentive reasons. It's hot. The Styrofoam ice is expensive and keeps getting damaged. It is a climate-emergency film, with a carefully crafted political message and a crew of dedicated creatives. They are determined to make it work.

This is the laughter of recognition. It is genuinely hilarious seeing other people focus their emotional energy when we know it is a self-delusional effort. We are all just going through the motions. Aren't we? We laugh at the indulgent actress playing all of her power games with the director who is trying to keep it together. Cassandra makes an appearance as the doom-monger climate-science adviser with some very sobering lines, which of course go unheeded.

So, when the ice and landscape stuff is cleared away at the end of the day, the fierce sun beats down on the director alone on stage. This play manages to court our attention by creating the world we work in, or the world we would like to work in. It teases our insecurities and mocks our ambitions, and then it melts away the ego considerations of the artist and finds a level of honesty that is deeply affecting. It preaches to us, us the converted, sitting here having paid good money for our tickets. We care, we are the good ones, but it doesn't let us off the hook.

Beth Wohl has written a very good play that bites us hard with its humour and its humanity. Of course we shouldn't laugh at this crisis, but actually maybe we should. The laughter makes you feel the truth of the world in your aching sides.

4 JERUSALEM
Jez Butterworth (2009)

'What the fuck do you think an English forest is for?'

It is almost impossible to write about *Jerusalem*. It is not just a play. *Jerusalem* is a lament for an England under threat: England as a green and pleasant land, where characters churn up the language till we smell the leaf-mould and loam. England in its constant moment of change and rebirth.

We're in a messy, magical encampment in the forest of a fictional Wiltshire village, where you could see the footprint of Jesus just as it springs back from the moss, fleeting, real, true, oath-worthy. Johnny 'Rooster' Byron is a leader, a storyteller, a conjurer of pagan magic. Written with fury and love, Jez Butterworth's play becomes England. It is ancient and new; it captures the moment when everything is about to disappear and yet can just about maintain its mythic strength. This is an England that will refuse to be subsumed despite the developers and their tarmac and their show homes.

The play does what theatre ought to do – to make it impossible for us to live without sensual richness, linguistic invention, electricity shot through the imagination; impossible to live an arid life, without dirt under your fingernails, and, in Rooster's case, weed in your nostrils.

The plot is simple. It is St George's Day and Rooster decides to take a stand against the ugly new housing development jack-hammering its way towards his forest. His caravan nestles under the canopy of the trees. Council officials come knocking to serve him with an eviction notice, but he and his band of merry outlaws refuse to budge. The stand-off that follows is a drink-and-drug-fuelled extravaganza of asserted English identity, protest and paganism. *Jerusalem* is a towering state-of-the-nation play about the radical heartbeat of England's countryside.

We are repeatedly told there is a desperate need for new housing stock, the shortfall apparently not matched by the number of existing homes that stand empty. Despite second- and third-homeowners buying up the countryside, new homes and mortgages are what our economic growth is built on, so we build more, and more. The losers in this capitalist bonanza are the edgelands around towns and villages which are absorbed by ever-expanding suburbs, their woods and scrub tarmacked over to become driveways and mini-roundabouts. Where are our children truly safest: in the woods, where they can't help but be their true selves, or on a modern housing development, all constraint and speed bumps?

Rooster, a towering leading role amongst a colourful ensemble, embodies the ancient Green Man of old pagan England, feral in mind and radical in spirit. Just imagine what would happen if we all danced to Rooster's drum.

MELT 5
Shane Mac an Bhaird (2019)

'To us. To humanity. To our unerring capacity for laying waste to the world while searching for impossible horizons. To destroying what we love. Good health.'

Grizzled old scientist Professor Boylan is in the freezing Antarctic, attempting to go deeper than any person has done before. He's on a mission to find a lake under the ice untouched by human interference for thousands of years. Along comes enthusiastic student researcher Cook, dreaming of glory and already practising his Nobel speech. They Skype back home to Boylan's ex-partner, Elaine. She runs the funding for this project and is dismissive of his chances of discovering anything of use. Cook has a crush on Elaine (she's his professor) and agrees to secretly report back to her about the worst of Boylan's dangerous eccentricities.

Fuelled by whiskey and ambition, Boylan is lowered into the ice. One hundred metres. Two hundred metres. A thousand. There's the lake! And there, on its banks, is a baby girl. Is she an illusion? A hallucination? Buried trauma made real? All Boylan can do is bring her up to the surface and see what happens.

This play starts as a seemingly straightforward drama about scientific machismo and the urge to go deeper/higher/further/faster. But it soon becomes a surreal and unexpectedly moving story about how wonder can turn into fear and how quickly fascination develops into destruction. The strange baby creature ages before our eyes, becoming a disruptive toddler throwing papers around the research hut, before moving into teenagehood and sexual maturity. After a somewhat bizarre sexual encounter with Cook, she nests and lays eggs before becoming an old woman and escaping into the howling blizzard outside the door.

As the storm gets worse and communication with the outside world becomes impossible, Boylan and Cook's supplies dwindle. They must decide whether to do everything in their power to safeguard the eggs of this extraordinary creature for posterity and scientific knowledge, or eat them to save their own lives. This strange, affecting play highlights humankind's folly in destroying what we love, and the damage we do to both ourselves and the planet in the process. The idea that this act of ultimate self-destruction takes place at the farthest reaches of the world evokes a quiet, lonely and sad outcome, perhaps an appropriately pathetic ending for our hubristic species.

6 TARAP MAN
Ann Lee (2007)

'I can guess what your little article is about: how fragile the ecosystem. How evil the company. How pitiful that everybody, but everybody, is screwing the environment.'

Investigative reporter Aashi has no time for climate journalism. With press freedoms in Malaysia restricted, she believes her duty is to domestic issues of pressing concern – deaths in police custody, treatment of migrant workers, prisoners on remand without trial – rather than the plight of the orangutan. But when she stumbles on a case of wrongful imprisonment, the question of natural freedom takes on a more complex hue.

The ever-expanding intersectionality of the climate crisis challenges playwrights to stake out fierce and hitherto unimagined battlegrounds. Ann Lee makes a shocking story of injustice into a pacy, thrilling play that asks hard questions of independent journalism and explores the mental-health impacts of incarceration. She asks, what happens to people's imaginations when they are shut away from the natural world? Destruction can happen inside as well as out.

Aashi and her colleague Leong Kim are frustrated at their editor Regina's cautious approach to publishing their stories. Increasingly forced to scrutinise instances of corruption in secret, the last thing they need is to have to train up graduate newbie, Cornelia. Aashi initially dismisses Cornelia's interest in the plight of Malaysia's coral reefs. But when it coincides with a new case of wrongful imprisonment, Aashi decides Cornelia can help with the heavy lifting. Together they search through decades of court records and legal transcripts, entice prison guards into revealing key information, and charm hospital nurses into offering up secrets. The journalists become galvanised to drive ever deeper into this newly urgent common cause. Cornelia proves a useful ally to the seasoned Aashi. They eventually track down Yew Chong Sze (the 'Tarap Man' of the title) who has been imprisoned for over fifty years – without trial or diagnosis of his mental health – since he was a child.

Tarap is a fruit found in Borneo. The prisoner earned his nickname by eating tarap when he was arrested erroneously for murder. Shut away from the natural world for decades, Sze's only comfort is watching fish in a tank, swimming through fake coral. Their aquatic imprisonment has become a mirror of his own, the wonder of the undersea world an escape from his lived reality. When Aashi comes face to face with Sze, she is unsettled to realise his years of imprisonment have left him with only a hazy understanding of his circumstances; his sense of time, place and language have all merged into a passive, hallucinatory state. In fragments and snatched images, he leads Aashi through the story of his lifelong imprisonment, his recollections of the nature from which he has been separated, and his understanding of the motives and attributes of humankind.

Aashi and Cornelia both lose their jobs; editor Regina fears the fallout from an exposé of the country's fraudulent judicial system and refuses to print their story. Inspired by Aashi and Cornelia, fellow journalist Leong Kim resigns in solidarity. They have reminded him that freedom is found in many forms – physical, mental and imaginative. And that a relationship with the natural world is at the intersection of all of their lives.

7 OUTLYING ISLANDS
David Greig (2002)

'If the environment is full of gamblers – savers will prosper. If the environment is full of savers – gamblers will prosper.'

Set in 1939, David Greig's play depicts a world on the brink of seismic change. This is a subtle play, witty and wise, an island-sized microcosm of mankind's planet-wide interventions.

Two bird-surveyors from the Ministry of Defence arrive on a remote, uninhabited Scottish island, home to thousands of breeding seabirds. There are nesting colonies of kittiwakes, guillemots, razorbills, puffins, fulmars and shags. Robert and John plan to camp for as long as it takes to do an inventory of the island's flourishing bird population. Farmer Kirk, the owner of the island, rows the men over in his boat, but doesn't reveal that this survey is anything but benign.

The Ministry of Defence, concerned about the coming war in Europe, want to test an experimental new weapon, anthrax. They want to learn about its destructive power by releasing it 'safely' on the island. They need to know how many birds are alive before the destruction begins. Kirk's daughter Ellen helps the two men undertake their survey. She witnesses Robert and John's horror as they realise their mission: '*They don't want us to observe. They want us to take a census of the living dead.*'

Remote islands, loved by naturalists for their unspoiled habitats and undomesticated wildlife, are also loved by those who need pristine baselines to conduct toxic experiments. From the Americans testing nuclear weapons in the Marshall Islands, to the Russian military's bio-chemical experiments on Vozrozhdeniya, the destruction of critical habitats in the name of 'preventative weaponry' is a century-long, governments-led catastrophuck. In David Greig's haunting play, we watch four ordinary people wrestle with the moral consequences of military recklessness. For landowner Kirk, the money he will get from the MOD is more substantial than any living he can eke out from the island. Concern for the birds does not carry enough weight in his decision-making, nor that the island will not support life again for many years. He argues that his life with Ellen needs supporting in the here and now, a very recognisable urgency.

The play links the behaviour of the island's bird inhabitants with its temporary human ones. Robert and John observe that some creatures live happily alongside one another, while others compete for food and resources. This allegory of the wider world as it marches towards conflict is drily observed. They also note the birds' rivalry for affection at breeding time, unaware that their own behaviour towards Ellen might be more revealing than they know.

In the twenty years since this play was written it has gathered a more potent punch. It allows us to see the complexity of the natural world with its delicacy and its human interplay, whilst making stark the insensate crassness of those who do not factor planetary health into their military calculus.

8 THE BREATHING HOLE
Colleen Murphy (2017)

Nattilingmiutut translation by Janet Tamalik McGrath

'A sea bear! They're ever so regal – wait till you see one – they're fantastically white and have black noses.'
'ᑕᕆᐅᕐᒥᐅᑕᑦ ᐊᖅᖤᐅᖬᑦ! ᑕᑯᕋᓐᓂᖅᑐᐸᓇᓗᐃᑦ—ᐊᖕᒋᖬᖀᓗᐃᑦ
ᕐᑯᖅᖬᖬᒃᑐᑎᒃ ᕿᕐᓇᕆᒃᑐᐸᓇᓗᕐᒥᒃ ᕿᖕᒪᖅᑲᖅᑐᑎᒃ.'

The star of *The Breathing Hole* is a one-eared bear called Angu'řuaq, whose presence haunts the stage like a ghost. The play is based on a traditional story and made in collaboration with Nunavut performing-arts societies and the English and Indigenous Theatres at Canada's National Arts Centre. The script is entirely bilingual, written in both English and Inuktut.

The play takes place over five hundred years. In 1535 an Inuk woman called Hummiktuq has a strange dream. The next morning she discovers a bear cub floating on the ice near a breathing hole. She risks her life to rescue him, adopts him and calls him Angu'řuaq. She teaches him to be kind and helpful, despite being told by members of her community that the bear won't be able 'to go against his nature'. But Angu'řuaq is no ordinary bear, this is a bear that can transcend time. Three hundred years later the Franklin expedition is attempting to find the Northwest Passage. The explorers are startled to come across Angu'řuaq, sensing he is special and yet terrified at the same time. Fast-forward another two hundred years and a luxury cruise ship is now making the most of the melting ice to traverse the now oily Northwest Passage. Circumpolar Oil rigs thump away in the background. Thinking there are no more polar bears left in the world, they set up a remote-controlled bear with her cubs for the tourists. And then they encounter Angu'řuaq in all his magnificence. The ship's crew are unable to cope with this 'real' bear and tragedy ensues.

This is a profound and satisfying saga set in an extreme northern landscape. All of the action takes place at the breathing hole, which is in Nunavut, the traditional homelands of the Nattilingmiut people. If this place were a country on its own, it would be the fifteenth largest country in the world. Colleen Murphy has a deep understanding of Indigenous culture and captures with heartbreaking delicacy the beauty and fragility of this environment. She stipulates that when the bear is on stage the audience should hear him breathing, that his sound should fill the theatre. This sound becomes inescapable; comforting, awe-inspiring and unsettling, until ultimately the breathing breaks our hearts, and the folly of mankind's quest for territory and resources seems never-ending.

Despite this, the play has hope at its heart. It was created through an exemplary collaborative working practice and translated sympathetically. It is today used for joint activism amongst theatre-makers in Canada. By celebrating the deep relationship the Inuit have with their environment and the animals they share it with, the play allows us to imagine what a thoughtful civilisation that lives with a spirit of cooperation and sharing might look like.

LANDMARKS 9
Nick Darke (1979)

'Tractors don't eat oats, Wilf.'

Nick Darke's play, set in 1930s rural England on the brink of mechanisation, is a funny and heartfelt ode to a slower way of life. Wilf loves ploughing the fields with his faithful old horse, Border Boxer. They can get through an acre a day. To Wilf it feels like real work, in tune with his animal. The time it takes, and the effort he makes, are matched by the pace and depth at which together they form furrows in the earth. Then his rascally brother-in-law Totty (who only last week sold him a sickly cow) suggests they go halves on a tractor. A Fordson. It's the latest thing, everyone else has one, and it'll take half the time to do twice the work. Wilf's daughter Alice isn't convinced.

The play is astute and moving about the tussle between farming with horses and the introduction of the tractor. With the consequences of industrial agriculture now well understood, *Landmarks* feels even more pertinent than when it was first written forty years ago. The Great Leap Forward in the 1930s and '40s saw the invention of powerful farm machinery and coincided with the Second World War. Many great technical endeavours of the twentieth century were sparked in this urgently inventive period. Naval blockades made food security a prime focus. Producing as much food as possible at home was the only goal. The subsidies and incentives, along with breakthroughs in technology and chemical fertilisers, meant the countryside suffered a toxic shock. Wild species were decimated leaving many today threatened with extinction. Nick Darke sites this play at the cusp of this critical step on that brave and (retrospectively) foolhardy transition.

The dialogue makes this play something special. Speech patterns are real and lived, the vocabulary effortlessly authentic. Vivid cameos of villagers animate the world. Mrs Mayse recounts tales of her (imagined) exotic holidays, describing the sights and sounds of places she will never visit. A stranger professing to be a birdwatching priest arrives, but he might, in fact, be a boggart, a rare creature of folklore, wise but with possibly malign intentions. Wilf and Alice are determined to keep a close eye on him. Nick Darke handles the encounters between his eccentric characters with grace and wit. The humour is genuine and affectionate, never elegiac or sentimental but reflecting both the beauty and brutality of farming life.

The second act takes place three years after the first, where hedges have now been removed to make room for the tractor, the noise of which has scared all the birds away. The tractor gets stuck in the mud. The only way of getting it out is to get the old horse, Border Boxer, to pull it out. This is a humane and compassionate play about the farmer's constant dilemma to grow more, or grow better.

PART 2
RESOURCES

Resources

Who cut down the last tree on Easter Island? Who condemned that ancient civilisation to total collapse? As one of Aesop's lesser-known fables warns us, when the woodman came into the forest to ask the trees to give him a handle for his axe, why did the trees say yes? Without trees, no insect, bird or animal life can flourish; no undergrowth can form; no shelters can be made; no fires can be lit. The mindset of whoever wandered up to that final tree, axe in hand, seems baffling to us today. But can we really say we've learnt that simple lesson?

Finitude is a surprisingly difficult concept. We tell our children to enjoy their sweets, because 'when they're gone, they're gone'. But really, we know the shop is full, the adult has money, the sweet factory has magic machines. Perhaps it's this mentality that proves so hard to shift when thinking about the environment. As human populations increase, rates of consumption follow. Increasing pressure is put on resources and the reserve stocks deplete. The problem is the Earth cannot work as fast as a factory in churning out more. The environment simply cannot keep pace.

In this chapter the plays interrogate our compulsion to take without replenishing. Whether that be the cobalt for our mobile phones in Adam Brace's *They Drink It in the Congo*, or the weaponisation of water in Sabrina Mahfouz's *A History of Water in the Middle East*.

These eight plays explore the arbitrary way geology influences the creation of national borders, creating both winners and losers in the race for resources.

10 THE CHEVIOT, THE STAG AND THE BLACK, BLACK OIL John McGrath (1973)

'It begins, I suppose, with 1746 – Culloden, and all that. The Highlands were in a bit of a mess.'

John McGrath's landmark play-with-songs is a raucously entertaining account of how Scotland's economic history is inextricably linked to its environmental history.

Half of Scotland is owned by five hundred people. First they came for the grazing, clearing crofts and common land in order to graze cheviot and later blackface sheep. Then they came for the bloodsports, creating hunting estates for the gentry, decimating the treescape as they did so. And then, in the '70s, they came for the black gold found under Scotland's seabed, which has powered its economy ever since. It is a nation built on resource exploitation, by initially external and laterally internal powers.

The Cheviot, the Stag and the Black, Black Oil is funny. It is a furious and inventive journey through the nation's history. Presented as a ceilidh, this theatrical extravaganza broke new ground in every aspect of its creation. It was collaboratively researched and rehearsed, leading to a democratic and generous performance style. The play mixes historical information with variety and popular entertainment, creating an inclusive and accessible evening out for audiences young and old, habitual theatregoers or not. Its initial production toured to social spaces and non-purpose-built venues across Scotland, taking popular political theatre to those who might not otherwise have had access. A totemic production when it was first made, *The Cheviot, the Stag and the Black, Black Oil* became a cultural phenomenon; much imitated, never bettered.

A new version of this play might deal with the salmon, the wind farm and the big bad wolf. Scotland today positions itself as a provider of renewable energy, inviting new international investors to create different kinds of offshore deals. Instead of sheep, now it relies on salmon, with huge fish and crustacean farms forming a new monoculture. The tree-stripped landscape remains, although planting and rewilding efforts are intensifying, accompanied in some instances by community buy-back of land. Changes are happening, but the historical imbalance of the minority owning the majority is a story still deep in the heart of Scotland, and repeated around the world.

There are more such stories waiting to be told.

11 A HISTORY OF WATER IN THE MIDDLE EAST Sabrina Mahfouz (2019)

'Water-poor does not mean it has all been spent, but that the wealth has been taken away.'

This is a genre-defying show taking the lecture-demo format and turning it into a gig. Sabrina Mahfouz mixes music, song and poetry to dynamic effect as she takes us on a defiant dash through the turbulent relationship between Britain and the Middle East. No one is spared. Mahfouz excoriates the imperialists whilst also naming and shaming those responsible for human rights abuses. This is rousing and impassioned writing, ferociously intelligent and theatrically thrilling.

The abstract notion that something publicly available should be privately owned seemed absurd until we entered the collective delusion of the last two hundred years. Since then, natural resources can be controlled, priced, bought and sold. Access to fresh water is used as leverage by nations that have it, over nations that don't. In these times of disrupted weather patterns and frequent droughts, water has also become a market commodity. Water futures traded, speculated, sold short, investors betting on how scant water might be in a particular place at a particular time. Marketeers get rich on water-risk, on the life-and-death struggle of populations literally gasping for this life-giving resource.

A History of Water in the Middle East captures the lawless spirit and wild-west approach to today's water crises. Some desert states are planning to build snow stores, whilst others are drawing up plans to tow icebergs to the Middle East. One of the funniest lyrics in the play describes Jordan's water shortage being solved by training an army of female plumbers: '*Superhero with a plunger.*'

The play also deals with water becoming increasingly weaponised, such as the bombing of wells, a particular new low, a regular tactic of warfare, and the control of aquifers as one key strategy of Israel's blockade of Gaza. Several governments are investing in desalination techniques in the hope of discovering the alchemy that turns sea water into Evian. (This technology is currently more ecologically destructive than restorative.) Mahfouz also touches on the role of drought that drives the humanitarian crisis in Syria, contrasting this with a businessman skiing on fake snow in Dubai. The Yemeni capital Sana'a is due to become the first city in the world to run out of water. The irony is all of this happening in the Garden of Eden, a part of the world once known as 'the fertile crescent'. None of this is lost on Mahfouz, who sings '*it's an old, old war.*' Future world empires may be built on water rather than oil.

STRAWBERRY FIELDS 12
Alecky Blythe (2005)

'Is it worth it? That's what you've got to ask yourself. Is a strawberry worth... all this?'

Amongst the first of Alecky Blythe's verbatim pieces, *Strawberry Fields* uses her innovative audio style to interrogate the cost of diversification in farming.

The play was inspired by a Herefordshire farmer who decided to turn hundreds of acres of prime farmland into polytunnels to grow strawberries. To some people they were an eyesore, to others a sensible business move. Labour provided by low-paid migrant workers and the ethics of eating strawberries out of season were amongst the more contentious issues.

Alecky Blythe recorded interviews with the people at the heart of the issue – residents, local councillors, supermarket shoppers, farming representatives and the strawberry harvest workers themselves. She then edited together excerpts from these real testimonies to create an engrossing piece of documentary theatre. Actors who perform the play listen to the interviews through earpieces and reproduce every line, vocal tic, sigh, pause and cough verbatim, replicating with integrity the detailed words and unspoken thoughts of the interviewees.

Strawberries are always in season somewhere in the world. Rather than buy something that's been air-freighted and cold-chilled and wrapped in plastic (all reflected in its price and taste) from halfway around the world, is it better to try and grow them locally for as much of the year as possible? But actually, given the infrastructure set-up, biodiversity loss and intensive irrigation, not to mention the travel of thousands of seasonal workers who journey to help bring in the harvest, is the carbon footprint of an English strawberry in September any worse than an Ecuadorian one in March? Welcome to the minefield of food miles.

Food is an emotive issue, culturally specific and globally sought, a battle between localism and globalism navigating charges of protectionism and worries of food security along the way. As Alecky Blythe proves, a little strawberry can be a surprisingly dramatic subject for a play.

We include *Strawberry Fields* here to offer inspiration for alternative methods of creatively responding to the multiple environmental issues we face. Marking moments of change in a community by listening to their experiences is a critical part of activism, and a place where theatre can play a direct and useful role.

13 THEY DRINK IT IN THE CONGO Adam Brace (2016)

'The European mind says oh yah yah yah, it is not simple, wave a hand in the air, nothing is simple. And with the other hand keep taking. Keep take take take.'

They Drink It in the Congo explores the tensions inherent in trying to do something good about something bad. Stef, a British woman who has witnessed first-hand the violent reality of life in the Democratic Republic of the Congo, plans to put on an arts festival in London to raise awareness. She discovers that trying to get funders, artists, aid agencies, PR gurus, the Congolese Diaspora and several government departments all to agree on the content, messaging and format is a minefield. Richly informative, stocked with jokes and accompanied by a live band, this is a play that eviscerates the West's complicity in the chaos in the DRC, the most dangerous country in the world, according to the UN.

Using a smart potted-history stand-up sequence, we learn that the first resource taken from the DRC was people. Then rubber. Then palm oil. Today it is coltan, tungsten, tin, cassiterite, all the minerals mined at great human cost to make our mobile phones ever faster, sleeker, and cheaper. Adam Brace reveals how the West's parasitic reliance on these industries creates the perfect conditions for repeated atrocities. So, Stef asks, is the response to set up humanitarian festivals in the UK in order to soothe our guilt, or to genuinely do something more interventionist? In a committee meeting, Stef wrestles with the competing agendas of organisations that variously support the victims of rape in warfare, alleviating child trafficking, improving conditions for mine-workers, and saving the apes. She refuses to see any of these efforts as futile, believing wholeheartedly that doing something, anything, is better than nothing.

Anne-Marie, a Congolese campaigner, warns that the event is in danger of putting '*white words into black mouths*' and Stef is repeatedly hit with charges of virtue signalling, hypocrisy and conscience-salving. It's only when she faces her own demons by admitting the terrors of the atrocities she has witnessed that she can really begin her journey of understanding.

This play refuses to look away. With vivid cameos, dialogue that switches between Lingala and English, shockingly violent flashbacks, and joyous breakout moments of song and dance, this is an entertaining theatrical roller-coaster. The play itself is an act of awareness raising, admitting that there are no neat solutions. Perhaps this formally frank, politically unbridled play allows the simple act of staging this content to be a way of bearing witness.

14 FUCK THE POLAR BEARS
Tanya Ronder (2015)

'I dry my hair at the same time as boiling the kettle and watching Strictly, *I use an electric blanket – how do I hate myself, let me count the ways.'*

We've all left the lights on. We've all failed to separate out the plastic film from the tub. We've all showered for slightly too long after a hard day at work. Does it really matter just this once? Can one person really make a difference? This is a hilarious play packed full of relatable guilt.

Gordon and Serena live a very comfortable life. Gordon is a successful energy contractor. His wife Serena is pursuing a career as a fitness and mindfulness instructor. Their Icelandic nanny Blundhilde, employed to look after their young daughter Rachel, seems to spend most of her time sorting the recycling into the correct bins. When Gordon is offered a promotion to become the CEO of a fracking company, he and Serena begin fantasising about a bigger house and all the trappings of the middle-class life they crave. Feeling generous, they ask Gordon's long-lost brother Clarence, a recovering addict, to paint the house for them before they put it on the market. But when daughter Rachel loses her toy polar bear the household begins to unravel, domestic appliances begin doing their own thing, faces appear in the paintwork, Blundhilde is behaving strangely, and Gordon starts seeing the toy polar bear everywhere.

This is a domestic farce that's laugh-out-loud funny whilst sneaking a few home truths past even the most defensive 'issue radar'. Blundhilde's pet hamster spinning in its wheel becomes the perfect metaphor for the challenges facing this particular family, and for the rest of us, trapped in the never-ending cycle of accumulating, recycling and upgrading. The play tips into anarchy, with Gordon now out of control, eggs from eco-activists smashing against the windows and the hamster running riot. Through the middle of all this walks little Rachel, in her polar bear outfit.

With complete licence to be silly, *Fuck the Polar Bears* is as madcap as its title. Yet it makes a serious point about how to mobilise restraint and anti-consumerism on a global scale. Otherwise we all just think, 'Well, if you won't, why should I?'

15 FIXER
Lydia Adetunji (2011)

'They come to Africa looking for bad things, horror film. I can give them the cinema.'

This is a taut and absorbing look at the geopolitics of Nigerian oil. Chuks is a Nigerian fixer. For a small fee, he can take Western journalists to the source of the story or get them an interview with the people who matter. For another small fee he can do the same for a rival reporter. When an oil pipeline blows up in the north of the country, the world's press as well as the oil company's PR department rush to the scene. Who will Chuks help to get control of the narrative?

Laurence is a young British reporter chasing his big break. Dave, an old hack from a rival British paper, is confident he'll get there first. Determined to stop both is Sara, the disaster management guru for the oil company. She will downplay the incident and spin it for all it is worth. Her colleague Jerome is yet to decide which side of the fence he's on. Chuks navigates through these forceful personalities while all of them try and create a coherent and convincing narrative for their paymasters. Why have activists targeted the pipeline and what caused it to explode?

Fixer is a refreshingly original drama that reminds us that energy is not a simple climate-change issue but a tangled mess of geopolitical obligations. Oil-rich nations, like Nigeria, must battle with the mixed blessing of finding themselves geographically atop a deposit of something many other people value. Internal factions can be as demanding and fractious as external alliances. It is complicated being the financial beneficiary of what could be a stewarded resource but is, in fact, a major financial reward.

Chuks, the fixer, has his own agenda. The story he really wants the journalists to focus on is polio, the reoccurrence of which is still keenly felt by many Nigerian families, including Chuk's daughter, despite it being officially eradicated. All the Western journalists want the oil story, maybe with an added encounter with a guerrilla group, just the things to burnish the credentials of an awards-worthy writer working for a fashionable news outlet. Chuks finds himself no longer the puppet master. Instead he has become the victim of a web of back-stabbing and double-dealing. His life is ultimately nothing more than collateral damage, 'legitimately' squandered in the name of the story.

16 AN ENEMY OF THE PEOPLE
Henrik Ibsen (1882)

'The majority is never right. Never, I tell you! That's one of these lies in society that no free and intelligent man can help rebelling against. Who are the people that make up the biggest proportion of the population – the intelligent ones or the fools?'

Henrik Ibsen's seminal story of one man, Thomas Stockmann, who is punished rather than rewarded for exposing an unpalatable truth, still has the power to shock. Two brothers, doctor Thomas and mayor Peter, stand at opposite ends of a council meeting room and at opposite ends of the moral spectrum. One knows the town's water supply is polluted, its toxicity responsible for illnesses amongst the spa visitors who flock to the area. The other does not want this to be public knowledge for fear of economic ruin for the hospitality industries that depend on the tourist dollar for survival (much like the mayor in *Jaws* who would prefer his town's beach tourists not to know a shark might be in the vicinity. What they don't know can't hurt them, right?) Using sibling rivalry as a microcosm for civic responsibility, *An Enemy of the People* is a political thriller that lets us recognise when we give moneyed might the means to crush the morally right.

There are many examples of resources-made-toxic across our wounded world, from the relatively small crime of repeated and unauthorised sewage discharge into our rivers, to industrial pollution 'accidents' happening unchecked on a large scale over a number of years. Toxins in water supplies, which affect plant and animal species as well as communities, are invariably and incrementally devastating. The legal obfuscations, negligence and corporate irresponsibility of the Flint water crisis is the subject of another of the plays in this collection.

Those, like Thomas Stockmann, who find out such inconvenient truths, must choose whether to blow the whistle or not, to be rejected, slandered, or even murdered for their courage, or to side quietly with the polluter. The risk is huge. Many corporations are already on a battle footing, prepared to do anything to cover up responsibility for fear of the double whammy of financial loss and subsequent compensation claims.

A more diffuse enemy of the people, if we are prepared to say it, is the community itself. Why should we care about a few unlucky people? We've got businesses to run, children to feed, a meritocracy to comfort us in our good fortune. But in our quieter moments we know we have a choice: uncomfortable ignorance or terrifying knowledge. Neither is attractive. The illusion that everything is okay is getting harder to maintain. Our levels of denial are less reliable daily, our low-level panic is rising. We might all like to think we're Thomas, but perhaps watching this play we recognise ourselves in Peter.

FORWARD 17
Chantal Bilodeau (2016)

'I don't care what the weather reports say.'
Beat.
'What do the weather reports say?'

Against a soundtrack of Norwegian electropop music, *Forward* is a series of postcards through time, from 1893 to 2013, following forty characters in snapshots of their day-to-day lives at different moments in Norway's resource-rich history. For the first part of its trading history, the whale industry flourished, glacial deposits created fertile soils, which enabled the perfect conditions to grow ample food, the fishing industry thrived, all was good. Then oil was discovered and this new industry became the country's main source of employment and wealth. The play shows how each generation passes the baton to the next, revealing how the choices made by one cannot help but influence the next.

Forward is the second play of the writer's Arctic Cycle, a series of eight plays exploring the impact the climate disaster is having on the eight countries of the Arctic Circle – Canada, Norway, Sweden, Finland, Greenland, Iceland, the USA and Russia. Each play is researched and developed in collaboration with artists and citizens of those countries and reflects their stories and concerns. We've chosen this play as an introduction to the project.

Forward is a series of short two-hander scenes. We meet two whalers, two farmers, two fishermen, two oil workers, two environmentalists... After each of these scenes we come back to the fortunes of the Norwegian explorer Nansen, the story that acts as the spine of the show. We follow him in the 1800s as he travels up to the 60th parallel, dangerously mesmerised by the shifting ice. 'Ice' in the play is a seductive singer, siren-like, calling him on.

For those nations who have built their power on expansionism, the ice remains virgin territory. Competing countries plant flags both under and upon Arctic ice to claim future access rights to sea passages once the ice is gone, and drilling rights for future oil fields. The play takes the long view, placing this in the context of a long history of resource exploitation, and making a recurring joke that each generation ignores the warning signs that their particular resource is not infinite. This is clean, economical storytelling in bite-sized pieces that lands its cumulative punch with glacial heft.

PART 3
CONTROL

Control

King Canute, raising his hand to the tides, might have been the first person to demonstrate the folly of trying to hold back the Earth's natural forces, but he certainly wasn't the last. Different religions, cultures and societies have long asserted that mankind is the winner in the evolutionary race and therefore must possess dominion over nature. Yet history is littered with the rubble of civilisations who proved themselves unable to come to terms with the idea that nature has its own agency. Can we in the modern age learn this before we are taught it?

The attempt to control is a powerful impulse. The scale of our ambitions uses nature as the competitive measure. From mammoth infrastructure projects to hold back the water, to nanotechnology and gene editing, it seems as if we just can't live in the world without redesigning it. The plays in this chapter tell stories about people who attempt to exert mastery over nature, from Prospero in Shakespeare's *The Tempest* unleashing a storm, to Claire in Susan Glaspell's *The Verge*, a botanist on the cusp of pushing a single plant through an evolutionary milestone.

Exerting dominion over others is also a powerful impulse. Control that manifests itself as the attempt by one nation to rule another, or to empire build, or to marshal huge industrial forces in the geopolitics of an already-plundered globe, is increasingly being recognised as the aggravator of the climate crisis it has always been. Developed nations imposing their style of consumption, their ambitions, and their grievances on other nations is a form of control as cultural erasure. Nations and peoples who have lived for thousands of years in reciprocal, low-impact manner, deeply connected to nature, are themselves now under threat of extinction. First Nations peoples are an increasingly powerful voice in the climate fight. Their knowledge of how to live sustainably, with a profound understanding of the concept of stewardship is core knowledge which may help get us safely into the future.

The plays in this chapter touch on the pleasures, rewards and dangers of trying to control our environment with a rich array of provocations.

TRANSLATIONS 18
Brian Friel (1980)

'We name a thing and – bang! – it leaps into existence!'

On first glance, cartography could seem like an esoteric, innocent activity. What could possibly be suspect about surveying a few fields, drawing a few lines and rewriting some old signs? And yet, Brian Friel's masterpiece reveals map-making to be a deeply political power play. Whose understanding of place is being transcribed? What sly cartographic grammar is being drawn? What is given prominence and what is erased? And, crucially, what language does this map speak?

Translations is set in 1833 in rural Donegal, Ireland. Owen is a local man employed to provide English translations of Irish place names for the Ordnance Survey's newly commissioned map. He's working alongside British soldier Yolland, and his hard-nosed boss Lancey. Yolland's romantic nature falls in love with the Irish landscape, while Lancey, the experienced cartographer, is savvy enough to know this is a mission to suppress. Yolland senses the linguistic vandalism that will result if they continue to anglicise Gaelic place names. A love triangle develops between Yolland, Owen's schoolteacher brother Manus, and Máire Chatach, an ambitious, progressive student who wants to learn English so she can emigrate to America.

Disregard of place and language has implications for the planet. There are many examples around the world where erasure of place-specific language has led to a subsequent loss of knowledge and practice. A name may reflect the spot where a particular plant thrives, or a critical rest point for grazing herds, or a place prone to flooding. Lose the specificity of knowledge locked safely inside the language and the conversation between the land and the people, accruing for generations, is wiped, a lost library greater than Alexandria. In this play, an area known as *Poll na gCaorach*, meaning 'hole where the sheep shelter' in Irish, is reduced to the meaningless Poolkerry in English. A central strategy of invading powers, keen to appropriate land and resources of sovereign states, is to destroy their culture. Languages, rituals, social practices and norms, all of these are undermined, subdued, attacked or erased, and all of these sever connection with the way the earth is asking to be worked.

Translations is remarkable for the trick of convincing us that multiple languages are being spoken on stage throughout, whilst we only hear English. Brian Friel ingeniously uses the gestures and body language of intimacy to make it abundantly clear who is speaking what language and who can understand. A dynamic theatricality is built on 'living' languages. Characters reach for connection, or at other moments pretend ignorance. A rightly famous love scene, in which neither character can understand what the other is saying, is exquisitely heart-rending and hilarious for an audience who understand both tongues.

As the drizzle (*ceobhràn*) comes down, Friel depicts a community whose nuanced understanding of its environment, with long histories of lived experience kept alive in its place names, is under threat of destruction forever.

When we finally get to the last day on this planet, this play may well be a good one to perform. It is the perfect eulogy, a song of how we erased ourselves. Practical knowledge and cultural identity are the most valuable 'commodities' we have. How about we put a price on them?

19 KATRINA: THE GIRL WHO WANTED HER NAME BACK Jason Tremblay (2016)

'They build them levees. They build them levees cause they know they gonna break. Give folks a false sense of security.'

This thoughtful play for young people asks what it means to survive and recover from a generation-defining tragedy. It is an exuberant actor-musician piece. Every step of Katrina's journey is followed by a live jazz band, playing through the pain, stirring up the hope, in classic, joyful, New Orleans style.

Tropical cyclone Katrina made landfall across the southern United States in August 2005. Katrina's strength was as high as the scale allowed right before it hit New Orleans. The levees built to protect the city, at a cost of billions, almost universally failed. 80% of the city was left under water. Officially, some 1,500 people lost their lives, although the actual figure is widely presumbed higher; countless others lost their homes and livelihoods.

Much has been written about why the New Orleans levees were breeched. From French and Spanish colonists in the seventeenth and eighteenth centuries ignoring the advice of Native Americans and building in all the wrong places, to modern authorities failing to implement flood-insurance schemes, to the city's black communities being historically under-resourced and underprotected; suffice to say it was unequivocally a failure of political leadership in both the event and its aftermath.

The play focuses on Katrina, a young girl separated from her father during the night of the hurricane that shares her name. She helps her elderly Aunt Beulah and her irritable wheelchair-bound neighbour Mr Thibeaux to set off for the safety of a local church. The trio get lost. They seek refuge in the closed-down Perseverance Hall, one of the oldest music venues in New Orleans, still haunted by the musical spirits of the city. In despair and rage at what is unfolding around her, Katrina shouts her name into the wind, shedding her association with this terrible storm. She encounters an apparition, in the form of a seventeenth-century ghost, Toussant, who helps Katrina see things very differently. Gradually her bravery returns and she determines to recover the magic within the city, the magic within herself, and in the process reclaim her name.

Katrina is a wonderful central character, a young woman who holds the entire narrative and emotional agency of the story. By turns cheerful, petulant, angry and finally wise, Katrina teaches her elderly companions how and why people choose hope.

'Adaptation' and 'resilience' are environmental management buzzwords about how communities might in the future be better prepared for living with the unpredictable effects of a climate in flux. A more sensible approach than the King Canute one, perhaps, but in danger of camouflaging many of the underlying social inequalities that leave certain people more vulnerable than others, a useful smokescreen for underinvestment. 'Resilience' of any community is predetermined by historical resource distribution. Environmental racism is a pernicious and overlooked aspect of disaster response. The wealth disparity in New Orleans left whole communities living in areas more susceptible to flooding than others. The African-American population was disproportionally impacted by the worst of the flooding. Neighbourhoods were destroyed, leading to the displacement of thousands of black citizens of the richest economy on Earth. Still those inequalities persist. Flawed decisions, based on habitual disregard, in both immediate and long-term aid distribution continue to impact those who stayed and the wider diaspora of that proud city population.

Katrina: The Girl Who Wanted Her Name Back celebrates the survival and indomitable spirit of one of America's most extraordinary cities.

20 THE VISITORS
Jane Harrison (2020)

'Visitors don't stay. They're visitors.'

This play pivots on the moment the First Fleet dropped its anchor in Australia in 1788. Let's rephrase that sentence. This play pivots on the moment a group of Indigenous leaders gather to debate the implications of the First Fleet dropping anchor in Australia in 1788. Jane Harrison's *The Visitors* turns our familiar view of history on its head, focusing not on the predatory colonisers but on the First Peoples. This radical narrative gesture shows how theatre can give voice to the unheard. By skilfully reforming accepted narratives into more inclusive truths, Harrison deftly challenges the idea that history belongs to those who wrote it down.

A group of seven Clan Leaders gather on a headland on a sweltering hot day in January 1788. They scan the horizon, watching first one, then two, then ten large ships approaching. White men have visited before, the Leaders know their behaviours, but this looks like something more. Why are there so many? What do they want? Should they be welcomed, as is the accepted tradition? Or should the Leaders begin preparations for a council of war? Each Leader comes from a different region and represents a different Clan. They have divergent experiences and expectations of white men, and cannot agree on their course of action. Will they be able to reach a consensus before the boats disembark?

The idea of '*terra nullius*' – barren land – was more than just a myth that the British told themselves. It became a lawful definition by which, in the eyes of the Crown, appropriation and conflict were legitimised. This fiction belied the fact that Australia had been inhabited for at least 60,000 years. It also became the legal device that controlled the globe.

With no visible evidence of what the British would recognise as agricultural practices, such as farms, herds or enclosures, assumptions were made that any Indigenous Peoples living in this place had no relationship to the land, and therefore were not 'owners'. The understanding that First Nations peoples had indeed managed the land through a complex system of fire control has been a shockingly recent penny-drop. These techniques created specific habitats for particular plants to flourish which were then harvested on rotational cycles. It also generated environments for animals to thrive, to be hunted as required. A huge aquaculture site of fish traps, channels and weirs has recently been added to UNESCO's list of World Heritage Sites, firmly putting to bed once and for all the stereotype of the nomadic hunter-gatherer.

The United Nations Council of Indigenous Peoples gives a platform to First Nations people from around the world. Many of these come from traditions of stewardship with regards to the land, working on long-term cycles to ensure adequate provision for their grandchildren and great-grandchildren. This knowledge, largely stored in oral rather than written traditions, is increasingly understood as valuable knowledge in the global brain trust. This is also seen as an important moment for colonising nations to take responsibility for the damage they have inflicted on peoples and places around the world.

In *The Visitors*, Jane Harrison stipulates the seven Leaders should wear suits to signify their high status. Of different ages, temperaments and experiences, they speak English to invite understanding. And they express an array of cultural perspectives celebrating the differences between their clans, ridiculing any sense that an Indigenous population is a homogenised whole. Walter, from the Eel Place Clan, takes the long view, thinking first and foremost of peace. Joseph, from Spear Clan, feels sorry for the people on the boats, whose quality of life is clearly pitiful compared to his. Lawrence is young and wonders what tools the visitors might have. Gordon's experience of seeing his father shot by a white man eighteen years ago makes him wonder if the greater nobility is in fighting or in welcoming the enemy. Consensus amongst this group is not going to come easy. By dramatising this diversity of argument, *The Visitors* is an enlightening and entertaining provocation. What if a shared history could tell us something important about a shared future?

21 THE PLAY OF THE WEATHER John Heywood (1533)

'Sende us wether close and temperate,
No sonne shyne, no frost, nor no wynde to blow.'

Ah, the weather. Once seen as a peculiarly British obsession, it's fair to say that the state of the weather is starting to dominate many people's daily conversations: 'freak' heatwaves, 'unusually torrential' rainfall, 'uncharacteristically frequent' storms… the vocabulary of the abnormal in weather reporting continues to cloud the alarming truth that 'normal' evaporated long ago.

The Play of the Weather asks us to consider the serious question, what weather would we ask for if we could?

Many people who do not accept the climate crisis often point to the weather as evidence. 'There was a hot summer like this in 1937.' 'It snowed in June in 1953.' 'The water came up that high in '62.' Such statements referring to isolated incidents confuse weather with climate. Weather refers to particular atmospheric conditions over a number of hours or days. Climate refers to long-term averages over long periods of time and space, and has both short- and long-term variations. Short-term might be single dramatic events, such as El Niño. Long-term might be the gradual increase in average rainfall across Europe.

Heywood's play, despite its archaisms, suddenly feels rather prescient. Gods and goddesses control the weather – Saturn looks after frost and snow, Phoebus the sun, Phoebe is the rain-maker and Eolus shifts the winds. They've fallen out and in doing so have brought topsy-turvy weather to the Earth, causing all manner of natural disasters. Jupiter steps in as peacemaker and, as part of the diplomatic process, employs a character called Merry Reporte to ask various citizens what they think the weather should be. The Merchant wants strong winds to propel ships across the seas. The Gentleman wants temperate weather to be able to hunt. The Ranger wants storms to bring down trees so he can sell more wood. The Water-miller wants rain; the Wind-miller wants none. The Laundress wants '*sun sun sun to dry her loads*'. Finally, a young boy enters, asking for snow because he enjoys '*snowballe fytes*' with his friends.

Similar arguments play out at climate summits. Some scientists want to shoot volcanic ash into the atmosphere, simulating what happens when major volcanoes erupt, allowing large ash clouds to cover much of the Earth, thereby reducing the global temperature. If we did that now, they argue, we'd cool the planet for a few years. But what about if not all countries want to be cooled? Some people rely on certain types of weather and seasonal variation for harvest, infrastructure, national ways of life. Perhaps not everybody wants the same weather. Who would decide what to put into the atmosphere and how would they control the areas it would cover? Who would do it? The UN? The G7? Would a global referendum sort this out?

The play becomes politicised as Jupiter debates not only which weather is best, but which of these members of society is more deserving, or useful, or important, than another. In the end, Jupiter decides not to rock the boat but to stick with the seasons; to please some people some of the time. The judgement is that the weather shall remain exactly as it is. The most radical moment comes when Jupiter admonishes all the petitioners for thinking only of what they want the weather to be, rather than considering the good of the commonwealth as a whole.

Heywood could never have imagined that the content of his satire would cease to be a metaphor and become such a literal moral issue five hundred years later.

22 INFINITY'S HOUSE
Ellen McLaughlin (1989)

'When you and I went into physics years ago, we might as well have been going into theology, it was completely esoteric. But now – we have the fate of the Earth in our hands.'

Many plays have been written about J. Robert Oppenheimer, the 'father of the atomic bomb'. We chose this one because it grounds his work and that of his fellow scientists in the wider context of human intervention in place.

The play is set in the same location in New Mexico across four different time periods. The first is in the early-modern period, where an elderly Native American from the South-West prays for guidance. The second is set in the 1850s, and follows a group of European refugees on the pioneer trail. They find the landscape to be different from the land of their dreams. The third is the 1870s, and follows a group of Chinese railway workers laying the tracks that will change the face of America forever. And finally, in 1945, Oppenheimer and his colleagues nervously await the first detonation of a nuclear test in history.

The spot was chosen by the American military for its apparently isolated and meaningless location. Unaware of each other in their separate eras, this group of settlers, industrialists and scientists sense the presence of an older moral authority in the area, but are unable to fathom its meaning. Finally, the Native American, referred to as 'The Indian' throughout, brings them all into the same awareness of their lasting impact on this place.

This sprawling, intelligent play uses epic time to answer epic questions. By foregrounding Oppenheimer's growing awareness that he is about to irrevocably change the course of human history and habitation forever, the characters from the earlier periods haunt this beautiful palimpsest landscape. They are the spectres of past attempts to impose human authority on a wild and powerful place. By converging four truths in this barren location, Ellen McLaughlin creates a desert dreamscape with an unsettling and hallucinatory quality. People pray, dream, die and bury their hopes in the earth. What is left of us when we are gone? As the successful test bomb explodes and the cheering begins, Oppenheimer comes face to face with the Indian, asking for forgiveness as he realises '*Now I am become Death, the destroyer of worlds.*'

This provocative play embraces the idea that the Earth has been inhabited by different eras of knowledge. The trope of the magical Indigenous character has become much derided in recent years, but the employment of a character from deep time, from a cultural understanding of time different from this digital world, is powerful and effective. Humanity's ceaseless ambition to control nature for its own use, contrasted with an understanding that the past held all of these powers but did not exploit them, is a sobering and powerful realisation.

GREEN DUCK 23
Fabián Miguel Diaz (2012)

Translated by Gwen MacKeith

'I like ducks, he says, that's why I don't eat them.'

Green Duck is an eco-*Romeo and Juliet*, a tragic love story about a young couple denied access to the two things that make them happy: nature and each other.

The play is set in a mountain community in Argentina. The Boy is black, the Girl is white; he's poor, she's rich. The mountain is polluted by something unidentified. It's harming the children in the area. A boy has just gone blind; another has died. Boy in Love has scaly hands and feet; his parents pray for his scabs to heal. Lonely Girl burns if her skin even glimpses the sun; her dad worries she may die. Neither is allowed out to roam in nature or swim in the river, which they both long to do and instinctively feel will be healing. This is a beautiful coming-of-age tale, told through poetic intercutting monologues, revealing the delicate and powerful connections between love and the natural world.

The Boy and his parents are employed by the Girl's father to work his land. Boy cleans out the pigs. He spots the Girl through her window and they embark on a playful exchange of looks and letters. As they mature this develops into a powerful and sensual connection. Boy is told he must grow up, he must be a strong man, use a machete, hack away at the long grass, eat meat. Girl is told she cannot go out, that the elements will harm her. Both just want the chance to make sense of their bodies in the world and find their own way.

The healing properties of the natural world are at last being recognised. Children are encouraged to play outside, patients are prescribed wild swimming and hiking, forests are places for meditation and green bathing. The connection between nature and our physical and mental health has never been so well understood. *Green Duck* captures the sensuality, playfulness and tactile joy to be found in the natural world; the nature cure that offers recovery and freedom.

Boy rears a green duck to give to Girl as a token of love. Girl breaks out of her room and they run outside together, feeling the barefoot sensation of grass for the first time. Their parents separate them, but freedom tastes good and they want more. One night they experience the ecstasy of water, a skein of ducks racing overhead as they feel exhilaration in each other's bodies. They feel weightless, high on love. But the ducks are shot by Girl's father and green duck's wings are clipped by Boy's father. The lovers' bodies are found by their parents the next morning, entwined in the reeds.

24 THE VERGE
Susan Glaspell (1921)

'You can't shoot him in here. It is not good for the plants.'

This extraordinary play may have finally found its moment. It was thought too experimental in 1921 at its first production, and has been crawling towards the critical light ever since. A female botanist who dedicates herself to an ecological breakthrough, kills the man she loves, and rejects her daughter for thinking too conventionally about romantic love, is a central character we don't meet often enough.

From Rachel Carson's seminal *Silent Spring* to Hindu Oumarou Ibrahim's interventions on behalf of the Indigenous Peoples of the

World, women have long been vocal about aspects of the climate emergency that would otherwise be overlooked. Women's voices and perspectives have increased in presence and volume throughout the twentieth century as their lives were the first to be hit by the crisis.

Ecofeminism is a broad and dynamic field that is finding a footing in the global debate. Parallels can be drawn between the attempts to control nature and territory, and the urge to oppress women (almost as if they are symptoms of the same mindset). Ecofeminists have led in examining where social justice, environmentalism and women's rights intersect.

Susan Glaspell's play can be seen as a prototype ecofeminist adventure, or as the play that dramatised 'mansplaining' before Rebecca Solnit was born. The central character, Claire, is simply a woman trying to be her dynamic self. Claire has to go to extraordinary lengths to allow her inner sense of self to unfurl to its full dimensions. Her husband, sister and daughter all try to define her as not-doing-womanly-things. Her lover knows that, in order for her to continue to evolve as a human, he will have to leave. A gun is produced. The greenhouse is locked. Tension mounts.

The drama is played out through the expert attention to a particular plant, in a specially constructed hothouse, carefully observing and modifying the plant so that it can blossom into a new species; a plant on the cusp of its own evolutionary moment. This greenhouse is a place of solace and sanity. The natural world is foregrounded. The pain of the planet and the pain of the self are as one. She is focused, like Darwin witnessing the finches on the Galapagos Islands.

This play captures powerfully what it is like to be a woman in a world seriously destabilised by male appetites. Groundbreaking in its expressionistic trust that a nascent female dynamism could find form, the power and prescience of the writing is undeniable.

25 A COOL DIP IN THE BARREN SAHARAN CRICK
Kia Corthron (2010)

'How many gallons does one person in Ethiopia use per day? Three! One person in the United Kingdom? Thirty-one! One person in the United States? One hundred and fifty-one!'

This rich and expansive play yokes together eco-crimes on different continents and asks if there is a common cause.

A dauntless Ethiopian student, Abebe, arrives in Marlyand, USA, full of eagerness for his twin passions of God and ecology. He lost his family back home to poisoned water and is here to spend a year studying why. He dreams of returning home to help his community. His host family in America, Pickle and HJ, are themselves refugees from Hurricane Katrina. Having suffered the trauma of storm flooding, they are now experiencing the privations of drought. Pickle lost her son, father and husband during Katrina and life in recovery is tough. Her daughter HJ fears for her mother's mental health. Abebe becomes balm and solace to these broken people and they, in turn, take on huge significance in his life.

The role faith may play in tackling the environmental crisis is brought into sharp focus in this play. There are people who lay the blame for much of the world's problems firmly at the door of Christianity; the Bible's assertion that man has 'dominion over nature' was – and still is – taken rather too literally by some societies. There are others who see the engagement of faith communities as critical to averting climate disaster. 70% of the world's population profess to follow some kind of belief system and many of those religions hold nature sacred in some form. It may well prove that such deeply held values can motivate behavioural change, and rebalance believers' understanding of custodianship. The Parliament of the World's Religions, which has met annually for over one hundred years, focused its 2021 dialogue on the environmental crisis. A range of topics were discussed, including Daoist temples switching to renewable energy, the concept of a Green Ramadan, Jainism's vegan principles, Pope Francis's support for international ecocide law, and the role of Hindus in cleaning the polluted sacred Ganges River.

For Abebe, Kia Corthron's charismatic central character, it is his vital and energised mix of faith and ecology that keep him buoyant in a precarious world. The play ranges through time. We flashback to Abebe's life in Ethiopia. He puts his ecological learning into action to improve the life of his village, only to see all his small-scale changes lost beneath a vast hydroelectric dam. We travel forward in time to America, eight years after we first met HJ and Pickle. We see the same water corporations operating with disregard for nature and communities in Abebe's adopted country too. A baptism in the river cannot be performed because the local water company has diverted the river and caused the crick to run dry. Connection between faith and its rituals is now profoundly and irreversibly ruptured, no longer a readily available aide to survival.

For those with faith and those with none, *A Cool Dip in the Barren Saharan Crick* tackles a vivid array of eco-threats and a winning belief in the goodness of humans to survive and repair.

THE TEMPEST 26
William Shakespeare (1610)

'Now would I give a thousand furlongs of sea for an acre of barren ground.'

The Tempest is one of Shakespeare's last – possibly his very last – play. It opens with the most iconic sea storm in theatre history. A ship is being battered by the elements and all souls are thought to be lost. The characters wash up separated and disorientated on an enchanted island. Spirits, creatures and apparitions confuse and terrify the bemused survivors. We meet Prospero and his daughter Miranda and discover the storm was conjured by Prospero himself. The spirit activity taking place is also apparently at his behest, implemented by his servants Ariel and Caliban. Has he the right to be in this place, controlling the natural elements to exact revenge for injustices he believes himself to have previously suffered? This becomes the central question of the play. In Act Five all the characters meet in one place and an offer of reconciliation is made. *The Tempest* holds an enduring – if rather uncomfortable – fascination as a play about the power of nature and the dangers of appropriating that power.

There are scientists who believe the climate crisis should not be tackled by reducing emissions or curbing consumption but instead by humans directly controlling the elements. Solar Radiation Management (SRM in geoengineering jargon) proposes that giant mirrors in the sky, reflecting the sun's rays away from Earth, would reduce the global temperature. Another idea is that firing rockets into clouds to make or disperse rain is also a sensible option. This weather modification technique, famously used by the Chinese government to prevent a washout at the 2008 Beijing Olympics, continues to gain international traction. If there is a character in world drama who exemplifies this kind of hubris, it is surely Prospero, a prototype climate scientist.

The image of Prospero and his baby daughter in a tiny boat on an infinite sea is as moving as it is political; a parent protecting their child from the harms of the world, two refugees relying on nature to provide safe passage as they flee persecution. But Prospero's initial gratitude to the elements is transformed into arrogance by the time we meet him at the play's start. That his retribution takes the form of conjuring a terrifying storm, in which his brother believes his own child to be drowned, is a malevolent exorcising of his own deepest fears.

From Prospero's rage comes other controlling behaviours, the claiming of the island for his own (which is often read as an indictment of colonialism) and the enslavement of two more-than-human creatures, Ariel (of the air) and Caliban (of the land). The relationship with these two beings further seduces our senses by connecting us with the earth in its fullest natural expression, encompassing as they do the feeling of flight, the power to create fire and Caliban's encyclopaedic knowledge of the flora and fauna of the island. Shakespeare contrasts this with Prospero's learning, found in the written word, a relentless pursuit of scientific power which ultimately only brings him unhappiness. In the final moment of the play, Prospero throws his books away, vowing to reject control, accept vulnerability and be a better human.

Do Ariel and Caliban get their true freedom? Will Miranda inherit a world more in tune with its natural heartbeat? Can art and nature conjoin to offer sympathetic ways of humans and non-humans facing the future together? Shakespeare, as with most things, might just focus these questions for us.

PART 4
ENERGY

Energy

A popular Saudi saying supposedly goes: '*My grandfather rode a camel. My father drove a Mercedes. I fly a jet plane. My son will ride a camel.*' Seen like this, the Age of Energy is a graspable time frame within the history of one family, yet we struggle to understand that it is a mere blip in the history of the Earth. Fossil fuels took millions of years to create and a comparative nanosecond to extract. These fuels have powered the last two centuries of human activity and will shortly run out. What then? Perhaps our beasts-of-burden have enjoyed not being the dominant mode of transport for a while. They may soon find themselves loaded up again.

'Peak oil' is the name given to the moment when the maximum rate of extraction and consumption is reached, after which the oil industry enters terminal decline. It is predicted to be sometime before 2030. Some observers believe it to have already happened. Perhaps the day the balance is finally tipped is today, as you are reading this. Similar industries have similar peaks, all hotly debated as to their imminent/already-peaking status. The shift to renewable energy (technologies that have existed for decades) has therefore started to accelerate, albeit at different paces in different countries. But here the issue becomes a dilemma: do we need to find new ways of generating the same amount of energy as now? Or should we focus instead on just using less?

The plays in this chapter – from Tom Wentworth's comedy about fuel poverty, *Windy Old Fossils*, to Ella Hickson's searing exposé of a global industry, *Oil* – all explore how the search for energy is not an isolated activity. It has human, social and market consequences. It causes environmental damage, impacts personal relationships, creates community pressures, and is the dynamic behind financial overaccumulation in developed countries, leading to extreme global inequalities in developing ones; a merry-go-round of geopolitical tensions. These plays wrestle with the scale of thinking we need for this energy conundrum.

OIL 27
Ella Hickson (2016)

'You see a straight line stretching from here – from where you are now – can you see – you – on the horizon, she's travelled so far, she's doing incredible things in the world – look at her, can you see her? She's looking back at you – she's smiling – she's saying "it's going to be ok" – "it's going to be better than you can possibly imagine".'

Ella Hickson's astonishing play follows the lives of a mother and daughter from a candlelit Cornwall in the late 1800s through an emergently powerful Persia (now Iran) in 1908, a money-grabbing London in the 1970s, and a devastated post-conflict Iraq in 2020, before landing back in a much-changed Cornwall in 2050, when the Chinese lunar-mining programme is in full swing.

Hickson's play tracks the parallels of energy imperialism and female emancipation. May is trying, exhaustively, to give her daughter, Amy, a better life. Her determination to take what she feels she is owed pits her against her daughter's simple wish for a life filled with love. By linking the exploitation of resources with the exploitation of women, the play explores multiple issues of control and culpability, audaciously charting the birth, peak and demise of Britain's energy imperialism, alongside the Thatcherite version of feminism that has influenced May. The inevitable kickback from daughter to mother is mirrored in the rebellion of oil-producing countries to their former colonial overlords.

The play ends by offering a tantalising glimpse of how the next few decades could unfold in the inevitable resource wars. As the geopolitical landscape enters a new phase and the Chinese become major players in both renewable and alternatively sourced fusion energy, resource wars are likely to play out along even more complex – but no less empirical – lines than in the past.

As provocative as it is ambitious, *Oil* is a thrilling play, simultaneously historical and futuristic, asking us to consider our troubled relationship with the world's most conflicted – and finite – resource.

28 THIS LAND
Siân Owen (2016)

'So it seemed that there was some kind of magic in this field. Something underneath. Some said there was a dragon that had been woken by the lightning.'

Young couple Bea and Joseph have just had a baby and things aren't easy. Already exhausted and arguing, their fraught lives are further unsettled when a fracking company arrives in their village ready to dig. *This Land* takes us through the history of one patch of land and considers the impact the pursuit of cheap energy has on those who live on that ground.

The play employs multiple time periods weaving through a series of very funny scenes. From a Roman bureaucrat negotiating to build a road through a farmer's field, to the agricultural world-changing invention of Jethro Tull's seed drill, culminating in two forensic archaeologists in 2225 excavating a rubbish dump and wondering what on earth a Nokia 3310 might have been. Throughout, we see Bea and Joseph grapple with new parenthood and what it means to be a custodian of the land for the coming generations.

Geology is a lottery that creates winners and losers. Some nations (Iceland) run on cheap, renewable, earth-made heat sources. Others (Brazil) dig extremely deep for the coal that keeps their factories running. *This Land* imbues the earth beneath our feet with a rich imaginative presence, a dragon waiting to be woken, who will shake out of its slumber benign or furious, depending on the intentions of the digger. This skilful depiction of the earth itself as a force to be reckoned with has a powerfully destabilising effect on the audience. If the ground beneath our feet is not stable, what is?

The play chooses to stop at moments of 'progress' in history where land usage significantly changed. In this way it builds up archaeological layers of story and creates a perpendicular narrative structure. 'What time is this place?' is a question the play constantly asks, revisiting the same patch of land over and over in history, pitting the deep past up against the near future. *This Land* asks the audience to imagine vertically, down into the earth, as well as backwards and forwards along a horizontal chronology. At the intersection of this axis we find ourselves here, now, this place, this land.

The play allows flexible casting. It could be performed by two actors or a large cast. Whether using rapid-fire doubling to emphasise lineage and ancestry, or an ensemble to emphasise the communal issue, the sequential skin-shedding of the people in this place carries both metaphorical weight and theatrical charge.

Siân Owen arrives at the idea that 'fracking is fracture' and that fractures can be, and should be, healed. At its heart, *This Land* explores how the Hippocratic principle of 'Do No Harm' can be applied to the earth we stand on, and in this lateral imaginative leap we could reap greater benefits than merely heating our homes.

THE BURNING GLASS 29
Charles Morgan (1954)

'We haven't developed our scientific capabilities at the same time as our spiritual and political qualities.'

A British scientist working at 'a centre for weather control and understanding' accidentally discovers a new way of capturing and directing solar energy. Christopher Terriford's knowledge could benefit mankind, but it could also be manipulated by those with negative intentions. He is determined to protect his discovery but he falls under direct pressure from the government to reveal his secret. He hides the code in the one place he thinks it will be safe forever – his wife's memory.

Originally written as an allegory about the misdirection of nuclear power, *The Burning Glass*'s central metaphor has renewed focus. The real-world tensions between scientists working to better understand the Earth's natural forces, and those scientists employed by energy companies to develop better ways to control those forces, are growing. The ethics of this play are fascinating. In an era of energy crisis, should intellectual copyright of an idea rest with the individual, the nation state, or the energy company who could outbid everyone for that knowledge?

How does the global community collectively benefit? The play presses the question: who owns the secrets of the world's future energy sources?

The Burning Glass pits the individual versus the state, with the prime minister himself turning up at Christopher's home to pressurise both him and his wife. Mary has been entrusted with the code. She expertly steps up to articulate the depth of the ethical dilemmas at hand. (She has some of the best lines in the play.) Her formidable intellect, intuitive understanding of her husband's moral dilemma, and her awareness of the PM's fondness for 'a clever girl in a skirt', allow her some witty, intellectual manoeuvres. When Christopher is kidnapped by nameless agents of an enemy state, Mary has to decide whether to honour her husband's wishes or give up the secret to secure his return.

The moral struggles of the inventor have been dramatised many times. In some ways *The Burning Glass* is a conventional drawing-room drama with echoes of John le Carré. But its superior ethical debate, central roles for women, and its resonance with contemporary political contexts makes it a play that is due for reappraisal.

30 THE OIL THIEF
Joyce Van Dyke (2009)

'Sometimes I think, look, the entire Oil Age will have lasted maybe a hundred and fifty years, it'll be thinner than a sheet of paper in the rock record, how can any of it really matter?'

This is an intriguing love triangle with a female geologist at its centre. Amy works for a multinational, reading the rock strata to identify new oil seams. Her partner Rex is an actor, mostly living inside Shakespeare's head, unaware of Amy's growing unease with her job. Amy falls for Aleksi, a young translator who works for her company. He adores her groundbreaking work. An unspoken understanding starts to build, permissive but dangerous, as Amy works out her own place in the world and asks where her heart truly lies.

The most startling image of this play is the jar of crude oil Amy keeps in her apartment. It is from her first successful drill. Sixty-five million years of ancient African land has been layered and condensed into liquid gold. All three characters in the play are fascinated and appalled by the jar and the role Amy has played in extracting it. It becomes the currency of love. The eternity of love and the eternity of the earth become part of this complex relationship. The characters are specks in the vast river of time, and like Shakespeare's actors (according to Rex) they too will soon disappear.

The offstage world in this play is the Niger Delta, exploited for decades by Western oil companies; it is a place where the flares never go out and can be seen from space. In finally facing up to the environmental damage she has enabled, Amy realises she cannot bring her obsession with geological time to bear on her own relationships. She is causing lasting emotional injury.

A relatively short play, *The Oil Thief* parallels the idea of eternal love and the idea of geological time, and in doing so illuminates both. As Amy says '*a rock is time; oil is time.*' Geological space-time exists alongside human, personal time, both of them mysterious and powerful, both needing our attention.

WINDY OLD FOSSILS 31
Tom Wentworth (2016)

'The best thing you can do with that contraption is pull it down. I've told you before it's useless. Fine, have thirty of the things in the middle of the sea, but here? What we need is something practical.'

Lizzie is in her seventies and lives alone in a dilapidated cottage. Her husband Alan has just died and her brother Ted, with whom she has a somewhat spiky relationship, comes to stay. Unable to afford the seemingly never-ending electricity bills, Lizzie and Ted embrace candlelight and gas cooking as their modus operandi, attempting cheerfully to make the best of things despite a growing, if unspoken, sense of unease as winter approaches. One day, Ted hits upon the solution. They will attempt to resurrect the old wind turbine that Alan bought a few years ago ('*he was an early adopter, but that just meant it was out of date before it got used*') and wire it up to a bicycle. Perhaps if they take turns in pedalling, they might just generate enough electricity to put the heating on for an hour or two. Who knows, they might even get some music playing again and cheer themselves up with a dance.

There aren't many plays with two older principal characters, and even fewer which ask for the characters to squabble over who keeps raiding the snack tin, sneak around in the dark trying to find the hidden Christmas presents, and bicker with their ageing sibling about whose turn it is to pedal. This is a very funny play for two skilled comic actors.

Fuel poverty is a dangerous and shamefully secret part of many people's experiences. The fear of mounting bills puts older people and those living in isolation at particular risk. Tom Wentworth explores this painful subject with wit and compassion. Ted and Lizzie work to keep each other's spirits up, fight like cat and dog, and ultimately comfort each other in the (literal) dark nights of the soul. There is high comedy and raw emotional truth here.

The play poses some directorial challenges: scenes take place in the dark, seasons turn and characters noticeably weaken, and there is an offstage wind turbine, bits of which need to find their way on stage at various points for 'maintenance'. There is also, for those who want to produce theatre in a carbon-neutral way, the neat device of a working exercise bike that 'powers' the electricity – and therefore the production.

32 URANIUM 235
Ewan MacColl (1949)

'A uranium atom has ninety-two protons and a hundred and forty-three neutrons. Who could afford a cast of that size?'

In typically anarchic Theatre Workshop style, an improvisatory political ensemble of which MacColl was a founding member, *Uranium 235* moves between scenes, skits, songs and comment to tell a history of scientific thinking. It jumps from the Greeks to the Romans, with a sprinkling of medieval and industrial action, before settling firmly in the twentieth century to focus on mankind's most reckless discovery – atomic science. This is an illustrated, part-dramatised history of resource exploitation complete with theatre in-jokes and actor mayhem.

The play enjoys a metatheatrical structure. The second half begins with an audience member (a plant) interrupting: '*All this awful speechifying! If I wanted to listen to that kind of thing I would go to the appropriate place for it, a lecture hall. The theatre is hardly the place for it.*' The cast on stage take this advice to heart and start auditioning for more interesting acts. Marie and Pierre Curie sing a duet about their laboratory work and the mineral resources they are investigating. Marie then sings a tortured solo lament about whether to study sodium or mercury. The Ringmaster steps in to let her know her 'act lacks popular appeal'.

Einstein walks in and asks for an audition. His act is limericks:

'*Der was ein young fräulein called Bright*
Who could travel much faster than light,
She started one day
In a relative way
Und came back ze previous night.'

Finally, a dance explains nuclear fission (it genuinely does) and then a scene efficiently dramatises the worst impacts of the Second World War. In this ghostly aftermath, the play becomes unexpectedly moving: '*Excuse me, miss. I wonder if you can help me. I'm looking for the road that leads to a good life.*'

The play shows signs of dating, yes, but its new ending (updated in the 1970s) details how nuclear power transitioned from a weapon of war to an energy source. It includes a damning list of subsequent reactor disasters, health crises and associated waste problems. We include the play here as a gesture of gratitude and solidarity to a collective, democratic, populist company who pioneered a performance style that might be very useful to us all going forward. Modern science and the ethical debates we need to be having require a populist form and forum. Perhaps a music-hall theatricality might just help us laugh at the crisis and, in doing so, help us make sense of it all.

33 CARBON CHRONICLES Footloose Community Arts (2008)

'For all the problems we've got solutions,
We can plasticate the desert and grow algae in the oceans,
Use carbon offsets, clean coal, carbon sequestration,
At least we'll save our bit of the population.'

Carbon Chronicles is a comic participatory opera for performing groups to elaborate/adapt/improvise as they choose. It contains a brief history of energy exploitation in musical form, with songs such as: 'Petra Oleum Call and Response', 'The Chorus of the Ancient Plankton' and, our particular favourite, 'Song of the Combustion Engine'.

The piece, created by an improvisation collection called Footloose Community Arts, has been performed by different groups across the UK. It uses music and jokes to provoke discussion about a fossil-fuel energy system, and suggests ideas for how to live in a low-carbon society. It is written so anyone can take part, any age, any cast size, with or without musical experience. It is ideal for youth and community groups with a pick-your-own structure. It offers numerous ways for people to get together and sing silly songs about the climate crisis. One memorable moment is a chorus of elemental beings singing the story of their evolution into the tiny creatures that one day will be dug up by oil barons. The play also draws interesting comparisons between coal and diamonds, both made from the same matter and both exploited in equally destructive ways. A competitive duet between a lump of carbon and a diamond perfectly illustrates the point.

For all its jocularity, the piece does not shy away from serious issues. It draws attention to the execution of Ken Saro-Wiwa for leading a campaign against the degradation of the land and waters of Ogoniland in Nigeria. Fifteen years after his death, the Ogoni people were awarded millions of dollars in compensation for the destruction and for their distress.

Carbon Chronicles ends with positivity, with those who have taken part invited to imagine for themselves a scenario that best expresses their hopes and fears for the future.

ENDLESS LIGHT 34
Sayan Kent (2011)

'See how it sparkles. Feel it. Feel the power within it. How deep is that black. Like space. Thick black hair. A raven's head. That is pure nature. And out of it burns life.'

Nadia, who speaks these lines in Sayan Kent's thriller, is an excellent climate villain. She is the owner of the Mount Vardisi opencast coal mine in Kolkata, West Bengal, India. At the beginning of the play, she stands atop the mountain, surveying the bright lights in all directions powered by energy from her coal. At the end, she is subsumed by sludge, dragged underneath the black-water spilling from her own mine, a victim of greed and hubris.

Loosely inspired by Rabindranath Tagore's *Red Oleanders*, the play follows two estranged sisters who were born on this sacred mountain but grew up with profoundly different attitudes to the environment. Nadia believes the Earth's resources are there for the taking. We're all doomed, so why not make the most of it while we can? Her younger sister Chandra is the opposite, a serious and diligent eco-activist.

Chandra and her colleagues Seth and Kaz are taking soil and water samples that will prove the level of poisonous material seeping into the ground is far beyond internationally agreed levels. On the day of a new explosion to expose the mountaintop, they break through the cordon disguised as maintenance workers and begin filming the destruction. When the intruder alarms start ringing, Chandra and her sister Nadia unexpectedly come face to face.

Opencast mining, which uses blasting to access surface mineral deposits, produces vast quantities of waste from both extraction and processing. Black, carcinogenic water is contained in huge slurries which contain unprocessed sulphides, a toxic landslide waiting to happen. The fictional mountain in the play is based on several real outcrops around the world, including in the UK, in whose lower environs live hundreds of people. Valley communities have no choice but to live under the threat. Mine owners cross their fingers. In pithy encounters written with bite and conviction, the play adroitly captures this knife-edge risk-taking.

Endless Light follows the sisters as they battle out their differences, first through diplomatic channels at the Kolkata climate summit, then between themselves, and finally on the mountaintop itself, on the night of a torrential rainstorm which destabilises the loose scree. As the mud begins to slide, the sisters unite to try to prevent the worst. But is it already too late?

35 ARCTIC OIL
Clare Duffy (2018)

'Of course I want to tear it down. I want to destroy it and all the others. I want to throw my body into the world and make something change... or die trying.'

On a remote Scottish island somewhere in the North Sea, Karen is helping her daughter Ella pack for a wedding on the mainland. Karen has agreed to stay behind to look after grandson Sammy, who is asleep in the room next door. Karen clocks that Ella's packing doesn't look like she's off to a posh do. She locks Ella in the bathroom, challenging her to admit what she is planning. Ella caves. She is joining a group of eco-activists on a boat, sailing into the Arctic to attack a new oil rig. What unfolds is a battle of wills in a locked-room pressure-cooker scenario, as the women argue over how best to protect their children from harm.

Ella wants to live a life that has meaning, and for her child to grow up knowing his mother tried to do something. Karen believes looking after your family starts at home, not out on the ocean protesting. The argument is like power chess, truth and obfuscation ebb and flow, the debate keeps motoring, keeps surprising. Clare Duffy writes eloquently, at times violently, always presenting a disturbing picture of two generations at loggerheads. Their love and concern for one another cannot steer them onto common ground.

This play is full of tension, exploring self-interest and activism, in light of both local and global concerns. It exposes the familial divisions created by generational perspectives on the climate emergency. It touches movingly on the mental health crises that fester unattended in families living with extreme environmental anxiety.

The play reaches its conclusion with two shocking revelations. Karen needs to have a double mastectomy. As Ella reels from that information, news comes through that the protest boat has been attacked in the harbour. Her boyfriend, Sammy's father, is amongst the injured. Ella has to make a decision. Who will she go to? A cruel choice is left hanging in an ambiguous ending.

PART 5
POLICY

Policy

What is it really like to be in the room where it happens? Since the first global climate conference in Geneva in 1979, a succession of panels, committees and summits have held frequent get-togethers to discuss the waning quality of our environment. They leave a blaze of successful legislation (such as the Montreal Protocol in 1987, which led to the phase-out of chlorofluorocarbons); a trail of unfulfilled promises (the disappointment of Kyoto in 1997 still cuts deep) and a bewildering array of acronyms (BINGO, GOOS, QELROs, REDD and TUNGO, amongst many others, see over), in their wake. Reaching political agreement is challenging, so achieving *any* cross-border cross-party international legislation on climate issues can feel like a minor miracle.

Consensus requires compromise. It would help if compromise wasn't such a dirty word. Hours of negotiations and diplomacy are required for all parties to agree on even the smallest change in a statement. The energy put into wordsmithing could power a small planet. It also requires goodwill, a genuine desire to lay the foundations for legislative change, and positive action. Finally, it requires enough honesty to openly assess whether or not past targets have been met, and whether the responsibility has been accepted or kicked into the long grass.

The following plays all interrogate what happens at the top table. What happens when scientists, politicians, researchers, businesses and activists all come at the same issue from a multitude of angles? How are minute nuances in policy picked apart? Why do certain issues rise up the agenda while others slide down?

From Natal'ya Vorozhbit's searing *The Grain Store*, examining one of the worst political decisions in world history, to David Finnegan's outrageously funny *Kill Climate Deniers*, which takes a parliamentary occupation as its starting point; these plays shed light on the characters we hope will have enough vision to fix the emergency, and on the pressures they feel from all sides.

BINGO – Business and Industry Non-Governmental Organisations
GOOS – Global Ocean Observation System
QELROs – Quantified Emissions Limitation and Reduction Commitments
REDD – Reducing Emissions from Deforestation and Forest Degradation
TUNGO – Trade-Related Non-Governmental Organisations

KILL CLIMATE DENIERS 36
David Finnegan (2017)

'First thing I got wrong was the title. You want to call your play something fun, something playful, something catchy. But maybe with this I was wrong.'

David Finnegan began researching this play, a black comedy about a group of eco-terrorists taking over the Australian parliament, with the help of a grant from a publicly funded arts institution. When word of this planned use of public money got out, the trolling began, escalating through social media to shock-jock radio, culminating in mainstream attacks from the Murdoch press. The deliberately provocative title did indeed provoke. The one-line plot summary did set alarm bells ringing. Public money? For this? Finnegan had the option of engaging in patient and earnest rebuttals of this public bombardment, but instead he chose to use the discourse as material, writing the argument about the play *into* the play. What results is a hilarious and outrageous evening's entertainment.

Climate denial is real. A quick internet search reveals there are serious-sounding people who say serious-sounding things about the impossibility of humans impacting the planet's climate. Many governments and businesses are happily in step with this approach. The 'denial machine' is a phrase coined by academics to describe the feedback loop infinitely reinforcing itself, offering a comfort blanket to free-marketeers, contrarian scientists and populist politicians. Climate denial as a political position is growing. It piggybacks on conspiracy theories (largely an American phenomenon), on internet hoaxes, and on all manner of 'useful idiot' obfuscations by media companies whose investment portfolios would be damaged by significant change. Social scientists have fun creating a set of terms for each tactic employed: 'cherry-picking' from the statistics, 'manufactured doubt', 'magnified minority voices', all adding up to the stubbornly held, proudly anti-intellectual, tribal, reactionary, default position of 'climate denial'. It is so much easier not to care.

The plot of *Kill Climate Deniers* plays with these techniques of elasticating truth. An environment minister is taken hostage and a violent, high-stakes shootout gets underway. The action is interspersed with TED Talk-style climate-crisis information, and metatheatrical commentary from the actors on the choices their characters are making. The harassment suffered by the writer in writing the show is turned into a shocking and gasp-inducing added strand. The charged discourse in Australian society about the climate and the divisive way this is stoked in the media informs the play's intensity and boldness. This play is also a short film, a monologue, a concept album (which you can download and listen to whilst going on a tourist tour of the Australian parliament building). It is an ever-evolving amorphous work that has a life of its own and will keep changing as the dialogue about the climate catastrophe develops. As a play, it won't be to everyone's taste, but its power is in its witty fearlessness.

Reading *Kill Climate Deniers* can be a chastening dispatch from the front lines, a provocation for programmers and commissioners. Finnegan faces head-on many of the issues in sponsorship, legal challenges, direct action and governmental critique that arts companies will have to wrestle with. How do we avoid the chilling effect in subsidised theatre? To whom are we ultimately responsible? The ministers who fund our artistic work on five-year parliamentary cycles? The corporations who misdirect our attention? Or the generations to come who will wonder why we didn't say and do what was needed?

37 THE HERETIC
Richard Bean (2011)

'It's a Betel Nut Tree. No saline tolerance. I planted it – you know all this – on the wash limit sixteen years ago, and it's thriving.'

Richard Bean's provocative play tweaks the nose of climate doom-mongers. It flips the climate narrative on its head, focusing on a scientist whose research shows that sea levels are not rising. Silenced by her university and receiving death threats from activists, Bean poses the challenging question, is Diane a heretic or a victim?

Two rival universities are vying for funding in the new commercial playing field created by the monetisation of scientific research. At Diane's university, there's a new intake of students ready to sharpen themselves into clear-thinking scientists. Diane's specialism is sea-level change. She has planted a salt intolerant plant on the Maldives to measure sea-level rises – or, as her research appears to show, the lack of them. She concludes that the climate-disaster narrative is wrong and catastrophising by others is creating a story that is being embellished by activists.

The forty-three members of AOSIS (the Alliance of Small Island States, a coalition of low-lying and island countries) might just have something to say about Diane's analysis. Their latest statement tells of their 'profound disappointment' at the 'lack of apparent ambition' from the international community to 'protect vulnerable countries, their peoples, culture, land and ecosystems.' They are expected to be a major and urgent voice at upcoming summits.

Nevertheless, in this once flippant perhaps now reactionary play, Diane's plant hasn't yet died, meaning ocean salt is not yet reaching its roots. She even finds an unlikely ally in new student Ben, who on the face of it is an ardent environmentalist. He gets drawn in to Diane's world view through his crush on her daughter Phoebe, who suffers from anorexia. Diane's life is equally complicated through her on-off relationship with Kevin, her department head, who doesn't want her to release her research until the competition for funding is settled. The death threats she is receiving from an anonymous environmental group are not being taken seriously by the campus security guard. Diane defiantly appears on *Newsnight* to counter the Maldives' president's claims that sea levels are rising and gets suspended from the university. Tensions, unlike Diane's sea-level graphs, rise. A frenzied denouement involving data being hacked, a kidnap attempt, and a heart attack, ultimately leads to an uneasy resolution.

38 LIKE THERE'S NO TOMORROW Belgrade Youth Theatre (2020)

With Justine Themen, Claire Procter and Liz Mytton

'You know, they didn't even ask for our opinion. But it's our future they're shaping.'

This is a powerful, progressive ensemble play created in collaboration with members of the Belgrade Youth Theatre in Coventry. Their anger and clear-thinking speaks volumes.

The play opens with a ritual libation of the earth in rural Zimbabwe. Asha leads the ceremony, a gift to honour the earth from a community who live in balance with nature. Suddenly we are ripped away from this thoughtful scene into a busy Midlands city. Little cracks are appearing on the ground, and twelve-year-old Maru thinks she understands why. She's been reading a story about a community who honour the earth properly and she senses it may be connected. The city mayoral candidate, Bobby Brunt, has big plans to concrete over the city-centre park, replacing it with a new shopping mall and a housing estate. He promises residents he will reopen the car factories and bring employment to the area. Maru, a passionate environmentalist, sees Brunt's growing fame for what it is – a populist politician promising the earth to people short of hope. She doesn't want the park to be concreted over; it's where she goes after school to play. As an asthmatic who struggles with noxious traffic fumes, it's the only place she can truly breathe.

Her family aren't much support – takeaways, online shopping and computer games are their interests. When Bobby Brunt does his door-to-door campaigning, they are the first to welcome him in. But the day of Maru's thirteenth birthday proves a turning point. Her parents redecorate her room as a surprise, with new mass-produced furniture and brightly coloured walls. They throw away her old much-loved things, including the story book she hadn't finished reading. Deeply upset, a giant crack appears in her bedroom wall before spreading across the city. News crews, T-shirt sellers and fly-tippers descend, all looking to capitalise on the disaster. As Brunt makes a televised speech in front of an angry and frightened throng of supporters, promising to pour as much concrete as it takes into the earth to fix it, Asha emerges from the crack speaking Shona, the language of her people, and chastising the people for not honouring the earth. She and Maru join forces to win over the crowd and write the end of the story.

Like There's No Tomorrow has a terrific central character. Maru's personal mission to stop her city being polluted by the twin evils of toxic fumes and populist politics is rousing. This play has the wisdom of a young community built in. They see clearly the links between unbridled capitalism and environmental damage. The company have poetically dramatised the path that draws people towards quick-fix solutions rather than patient strategic outcomes. This is a call to arms from a young generation, an urgent and insightful play that speaks with an eloquent community voice.

THE GREAT IMMENSITY 39
Steve Cosson and Michael Friedman (2012)

'I guess I'm just not optimistic
I guess I look at a lot of statistics
But statistics can just leave you numb
Because statistics are deaf and statistics are dumb...'

The Great Immensity is a play with music: offbeat, accessible and entertaining. Karl was last seen on a tropical island near Panama whilst working as a film-maker for a nature programme. He's gone missing, apparently running off with a young eco-activist. His wife Phyllis deems the affair nonsense, and determines to find him. Through her search she uncovers a plot to destabilise the upcoming international climate summit in Paris. As the first day of the summit creeps ever nearer, Phyllis must enter the shadowy world of political surveillance, figure out what the hell is going on, and stop it happening before the entire summit is undermined.

The Great Immensity is an optimistic play. It looks at how hope for a better world drives both activism and science. It sharpens hope by pitching it against the alternative point of view that is determined to thwart altruistic idealism.

Executives from big businesses are increasingly present at the climate-talk table. They profess a desire to learn, they plead for time to change their working practices, and assertively press to have their voices heard. But some environmentalists see their presence as a hindrance. Stalling tactics are cleverly employed, offering false solutions, misrepresenting concrete targets, and misreporting progress. Campaign groups are often left trying to prevent executives with links to big agriculture and petrochemical businesses from attending multinational summits, preferring the seats to be given to those living and working on the front line.

The writer, Steve Cosson, and composer, Michael Friedman, spent time with scientists and nature documentary-makers to research this play, asking questions about what gives them hope. The play looks at how those on the policy front line cope with witnessing the tactics of vested interests. What makes them get up every day, collate data, record sightings and spend the laborious hours it takes to turn that work into clear narratives that politicians at climate summits can digest and act upon? The result is an uplifting Brechtian sequence of scenes and songs, expressing inner doubts alongside political clout.

40 E=MC² Hallie Flanagan Davies (1948)

'Today I put into your hands power OVER matter; the power of a – of a God, to slay and to make alive. It means food, shelter, abundance, for everyone.'

Ripe for an inventive revival, *E=MC²* is the history of the atom and how it transformed American foreign policy into a post-war fever dream. The tone lies somewhere between *Our Town* and *Dr Strangelove* written forcefully out of a witty, theatrical imagination.

*E=MC*² knits philosophy, science, and poetry together to dramatise how the atom was split. Flanagan Davies convincingly suggests that splitting the atom fractured human thinking ever since. The epic quality of the storytelling is genuinely exciting. She understands the grandeur of this mess we are in. She takes us back in time, beyond the Greeks, and also projects forward into future catastrophe. Part allegory and part skit, Atom is herself a character who explains to the audience just what splitting means, physically, scientifically and metaphorically.

Hallie Flanagan Davies was as pioneering as her plays. In the 1920s she travelled around Europe, studying the arts and falling in particular love with Russian theatre (Stanislavsky was a personal friend). Back home, in Depression-era America, she was appointed the national director of the Federal Theatre Project. The theatre industry had been hit by the double whammy of the Depression and the invention of the talkies. HFD stepped in to create a network of theatres nationwide that employed over 100,000 people making theatre in schools, hospitals, circus tents and factories, reaching millions of audience members with new plays.

One of her many innovations was the Living Newspaper Project, involving individual cities presenting dramatised versions of issues important to that community. The plays covered everything from racism to slum housing, on occasion leading directly to local policy changes. The form of the Living Newspaper was a montage of press reports, trial transcripts, fictionalised accounts and political reports, allowing audiences to witness social change as it happened. *E=MC*² is a stand-out example of one of these plays.

HFD was called to testify in front of the House Un-American Activities Committee, accused of supporting a communist agenda with her work. Maybe she was witnessing first-hand how domestic policy also turned into an atomic fever dream. She quietly moved into education, where her theatre teaching inspired thousands.

*E=MC*² is a useful outlier, reminding us that plays driven by nuance and subtext alone will not necessarily help us think our way out of this problem. We need epic, ambitious, thought experiments like this.

41 THE CONTINGENCY PLAN Steve Waters (2009)

'I just mean you could suggest that from now on climate-change policy is less about heading off the coming storm and more about weathering its affects.'

The Contingency Plan is two plays, either performed separately or together. They are set in the near future, dealing with changes and repercussions of British environmental policy over the last forty years. The turning point deftly captured by this intelligent writing is the shift we are making from taking all necessary preventative measures to asking instead, how do we cope with the aftermath?

Part One is *The Beach*. We meet Will Saxton and his dad. Will is a glaciologist, just returned from the Antarctic with new research on melting ice and its impact on coastal flooding. He's got a new girlfriend, civil servant Sarika, and he plans to join her working in politics. He feels his knowledge can be more useful there. His retired dad Robin is caught between pride and shame. He was once a policy adviser to the Thatcher government but failed to get ministers to take his data about climate change seriously. He now lives on the Norfolk coast with wife Jenny, deliberately letting the sea defences rot. What the hell does it matter now? He wants a front-line view of the rising sea level.

Part Two is *Resilience*, and is set a year or two later. Will is now advising the government and warns them that a massive storm approaching the UK will create a tidal surge. This will leave communities along the east coast under water. It's not what the government wants to hear. Another adviser, Colin Jenks, who was a colleague of Will's dad back in the day, dismisses Will's fears, just as he dismissed Robin's all those years ago. The storm hits and Number 10 goes into crisis mode. Does the government listen to Will and order mass evacuations? Or does it play the crisis down and go with Colin's gently-gently approach? Will desperately tries to get through to his dad to persuade him to leave the family home before the waves hit.

Steve Waters' precisely observed double-bill explores how those in authority arrive at policy decisions. Advisers offer best- and worst-case scenarios, with simulated likely outcomes, and advice on next steps. It is politicians who must decide what action to take. Should they let a disaster hit, and concentrate on the emergency response? Or take pre-emptive action and risk causing mass panic? The gripping situation-room setting of *Resilience* shows us what it's like to be at the heart of power during an environmental catastrophe.

THE GRAIN STORE 42
Natal'ya Vorozhbit (2009)

Translated by Sasha Dugdale

'Go and quietly give orders in the kitchen for everything to be taken back. Lock it all up and get it guarded. Arsei, Yukim, stand by the doors and don't let anyone in. I'll make a speech.'

The Grain Store is a Brechtian black comedy about how quickly one of the most fertile places on Earth can become a place of famine and forced exile. It is the story of the Holodomor, as visited on a small, close-knit village in Soviet Ukraine between 1929 and 1933. The Holodomor is a contested genocide, estimated by the UN to have killed up to 3.5 million people with another one million having fled. Various calculations have set the level of lost fertility at a further million, as those families still skeletal and diseased became the end-stop of their family line.

Natal'ya Vorozhbit's central character is a pyramid of grain which rots as the evening progresses. This is the fecund 'bread basket' of Ukraine. It is Kent with mountains. Stalin's first five-year plan was to strip the farmers away from their small-holdings and create vast armies of day-labourers on requisitioned land. Their knowledge of weather and soil, their local customs built on the seasons, and the pride and resilience of the farming communities were all wiped out. *The Grain Store* is a searing testimony on the colossal failure of the State to understand the people, and their relationship to nature. Husbandry versus Totalitarianism. This is a step-by-step guide on how to create famine, not avoid it.

The play centres on young lovers Arsei and Mokrina. They want to marry but are from different classes, Arsei is a peasant and Mokrina is a kulak, a member of the landowning class. The form and structure is where Vorozhbit's genius shines. An agitprop troupe perform a sequence of plays-within-plays. The tension grows between what is happening in the reality of the play and the false narrative of the costumed propaganda. The troupe foretell the arrival of immense political change, a vivid primary-colour joke at first that turns rapidly to a bleached-out, ragged nightmare. Arsei becomes employed by the Communists and ascends the ranks of the Soviet regime, while Mokrina and her family are stripped of their lands and possessions in the name of Collectivism.

Act Two of *The Grain Store*, set a few years later, explores how the genocide was hidden in plain sight. A *New York Times* reporter spends a day watching a Soviet film being made. During the shoot, villagers dressed up for the day in bright costumes, are asked to dance, again and again, take after take, dancing towards the tables laden with food, skeletal beneath their smiles and country smocks, never once allowed to eat. The reporter fails to see the dance of death unfolding before him. The world looked away.

PART 6
MIGRATION

Migration

'Climate change has wrecked everything; our people are living in other towns and cities, like refugees. All I wanted was to grow old with my children and their children. But now they are gone and I don't think they will ever return.' Shamisur Gazi, Chakbara, Bangladesh, 2013

The emotional rupture caused by the enforced movement of peoples has made this a painful chapter to write and assemble. To be ripped away from your family, your friends, your country people, and step into exile, is a living death, a ghost life. When we watch people walking over the horizon towards us, we should acknowledge this.

Disasters and wars linked to environmental issues are anticipated to displace 1.2 billion people by 2050. Human displacement can take many forms; a flooded Cumbrian bungalow to a church hall camp bed, poverty brought on by drought in an Indian village to the promise of a sprawling urban existence, the inexorable eviction of Indigenous Peoples, resource-war empty-handed flight. People can also experience displacement within their own location, an emotional rather than physical nomadism. A hurricane, a burnt-out block of apartments, dying crops from encroaching deserts – these events alter place-perception resulting in distress and disorientation within hearts, minds and communities.

The climate crisis is particularly adept at ripping people out of their environments. Lives have to be remade, futures reimagined, cherished potential abandoned. Becoming rootless in the midst of a crisis destroying your home is an estrangement from self, a slow-motion physical and emotional catastrophe. We have to open our arms.

Theatre understands place. Theatre is the place to understand exile. Theatre is where a sense of self can be made whole again. We have to give a platform to the lived experience of displacement on our world stages.

The plays that follow address migration in its many forms, from the unwanted odyssey of Nilo Cruz's *A Bicycle Country* to the slow-burn sadness of Wole Soyinka's *The Swamp Dwellers*, in which a subsistence family is undone by the toxic allure of the big city.

We all need the home this planet offers us. We all need our ancestors and our progeny to connect us to the geography that made us. The winds of the climate emergency will blow billions of us asunder. Theatre knows the vocabulary of welcome, the vocabulary of inclusion, and the vocabulary of home. We need to make company our mission.

A BICYCLE COUNTRY 43
Nilo Cruz (1995)

'We're in the Bicycle Age out there. We've gone back to the wheel. A whole country riding bicycles. You only have to look outside the window and see for yourself. Everywhere signs, slogans: save energy. What energy is there to be saved, when there is no energy!'

A *Bicycle Country* is a poetic look at the moment we feel we have no choice but to leave. It speaks to climate migrants all over the world setting off on perilous journeys across the oceans.

Act One is called 'Tierra' and takes place in a shabby hut on the outskirts of Havana. Julio has had a stroke and is housebound. He is looked after by his no-nonsense carer, Ines. She doesn't let him indulge his infirmity for a second and determines to steer him towards recovery. Julio's friend Pepe is a frequent visitor. He is a postman who rides a bicycle and always has news of how the outside world is faring. The three friends tease and berate each other, the dialogue crackles with as much energy and humour as the Latin music and dance they all love. But life is getting worse and they know it. Food shortages, power cuts, rising prices. They plan their escape.

Act Two is called 'Agua' and takes place on a raft in the middle of the Gulf of Mexico. They are hoping the currents will take them to America, where they will find safety and be able to live a better life. This refugees' escape is not the adventure they imagined. Exhausted, fractious and dehydrated, the characters begin to hallucinate. The play imperceptibly shifts into magic realism. Nilo Cruz's delicate writing distils their extremis into linguistic impressionism. They sing with defiance and grace, knowing deep down that all is lost.

The world is experiencing a paradigm shift in the movement of human populations. Whether people are escaping drought, floods, energy and food shortages or conflict driven by resource disputes, more and more people year on year are forced into exodus. New transit routes cut across shipping lanes, smugglers' routes and impassable rivers. The three main directions of travel are from South to North America, the Middle East and North Africa to Europe, and the Pacific Islands to mainland Asia.

This play is a simple three-hander that delicately pinpoints the nexus of decisions that lead people to leave. It also captures the ache of realising the better life on the horizon is always just out of reach. The wit and linguistic fireworks of the first half gradually shrink into the silence and emptiness of the second, a theatrical gear-change that makes for a sobering and absorbing drama.

44 THE SWAMP DWELLERS
Wole Soyinka (1958)

'Not a grain was saved, not one tuber in the soil. And what the flood left behind was poisoned by oil in the swamp water.'

Wole Soyinka's drama tells the story of a couple whose twin sons are forced to move to the city to provide a living for their family, only to find themselves adrift from their land, their culture, and their true selves.

Parents Alu and Makuri live in a stilt-house in a remote village in the Niger Delta. Flooding has destroyed the village crop fields, leaving households facing destitution. Makuri has an income as the community's barber, but it's not a lot. Their twin sons, Awuchike and Igweze, left to live in the city years ago. Awuchike has not returned for a decade, despite rumours that he made money and did well for himself. Alu fears he is dead. Why else would they not have heard from him? Igweze is expected back to check on the state of this year's harvest any minute.

The play begins when a blind beggar arrives at Alu and Makuri's door. He is from the north of Nigeria, where his village has the opposite problem to flooding. It is a drought-inflicted region. Alu and Makuri are spiritual people, they abide by the traditions of the village, including honouring the Serpent of the Swamp. This village deity guarantees clement weather and good harvests, as long as people pay adequate respect and fees to the Serpent's human envoy, priest Kadiye. He encourages villagers to sacrifice their best produce to him. The blind beggar believes in Allah, and becomes suspicious of the fat-sound to Kadiye's voice.

Igweze returns home, with painful news of his brother. Awuchike did, in fact, make a fortune in the timber industry. He sold wood to foreigners, little thinking that his new urban industry would indirectly contribute to his parents' demise. And now he has run off with Igweze's wife. The returning Igweze's fortunes are the opposite of his twin. He didn't want to make money in the same way as his brother and found himself isolated and in debt. Now home, Igweze's feelings break the surface. He lashes out at the life he has been forced to live in the city, pressured to make money, only to lose his wife and his brother. Now he has come home to find the harvest has failed. Igweze can now see through the religious hypocrisy of Kadiye. He now knows that the changing weather patterns are the forces of nature. The Serpent of the Swamp is nothing more than a story, told by Kadiye so he can get fat off his tributes. Alu and Makuri are shocked at their son's disrespect. They are challenged to defend the values of their small community.

This drama is a controlled explosion. Its elegant and contained storytelling gathers huge emotional power. The single-room, real-time action creates a breathless atmosphere, the tension slowly ratcheting up. Religious and spiritual beliefs are re-evaluated across a single generation, the amount of time it's taken for Nature's rage to near its peak. Igweze is left alone and heartbroken, only a stranger, a blind beggar, vowing to support him. His beliefs, his family and his home no longer have a place for him. Where does he belong?

THE GOLDEN DRAGON 45
Roland Schimmelpfennig (2009)

Translated by David Tushingham

'In the kitchen of the Thai/Chinese/Vietnamese Restaurant, The Golden Dragon: it's cramped, very cramped, no room, and still there are five Asian cooks working there. One of them's got a toothache: the Kid who's looking for his sister. The new Kid.'

The play opens in the kitchen of a restaurant with a cook, pliers in hand, about to pull out the tooth of the Kid. Out front, two stewardesses wait to order their food. Above the restaurant are many apartments, next door is a corner shop. We see into these lives, lived cheek-by-jowl in this crowded city. A girl is visiting her grandfather, a young married couple hit the rocks, loneliness creeps into the bones of the single man. They are all separate but connected in subtle ways, and they all at different times order food from The Golden Dragon.

Five actors play sixteen characters. There doesn't have to be any scenery, any costuming. It costs next to nothing to stage this play. The convention of doubling is made transparent. At the start of each scene the actors tell us who they are playing, where they are and what time it is. The five actors play all genders, ages and across a range of ethnicities. This play is visceral, moving and indelible. While your mouth is watering (they list the ingredients they are cooking, serving and delivering, it's hard to resist), the man reaches for the pliers. We are told that the Kid is screaming in pain. Our empathy is heightened, our mirror neurons twitching at maximum. We feel the pain and desperation of the young migrant working in the kitchen of this restaurant with all of our heightened senses.

The play was written for Berlin but could be set in any city. The playwright, Roland Schimmelpfennig, was inspired to write *The Golden Dragon* by wondering how undocumented workers with no healthcare and no papers, people in constant fear of discovery, live moment by moment. The movement of peoples across the globe is captured in a miniaturist snapshot of the action taking place inside a tiny kitchen. And yet the play also gracefully expands to be about the whole globe in motion.

This is a gateway play to the power of theatre, reminding us of how potent stories can be, how strong a fully engaged imagination can be. *The Golden Dragon* is a tiny miracle of a play; funny, precise and beautifully simple. Like a fable, there is beauty in the unadorned human response to this tragic situation.

LAMPEDUSA 46
Anders Lustgarten (2015)

'This is where the world began. This was Caesar's highway. Hannibal's road to glory. These were the trading routes of the Phoenicians and the Carthaginians, the Ottomans and the Byzantines... we all come from the sea and back to the sea we will go. The Mediterranean gave birth to the world.'

Anders Lustgarten's righteously angry play blazes through reductive media headlines to shine a humane light on the individual stories behind the silent migration, the one we choose not to see or hear.

Stefano used to be a fisherman, happy on his boat, working his catch. Now he is a fisher of men, pulling bodies from the same sea that once provided his livelihood. There's more money in people than fish, these days.

Denise is a British-Chinese student in Leeds. She's funding her way though university by debt-collecting. Forever an outsider, scratching a survival, she's doing what she can to give herself a chance.

Lampedusa is made up of intercutting monologues between these two characters. Their stories might seem unrelated at first, but gradually the links between migration and austerity, life and livelihoods, and the universal experience of poverty becomes clear. Our modern populist politics, that pits a nation's poor against its immigrant arrivals, is sharply realised here. Lustgarten makes the point that to divorce economic crises from humanitarian ones is to divorce wet from water. To further divorce humanitarian crises from its environmental roots is pernicious.

There is a brutal circularity to the environmental stresses that create movement of peoples. Fleeing parts of the world to then be attacked seeking refuge in the very nations responsible for their exodus is a cruelty of cosmic dimensions. Why are the questions about our moral obligation to refugees always whispered? The joined-up internationalist worldview dramatised in *Lampedusa* is characteristic of Anders Lustgarten's playwriting. Through these two amenable characters, the warmth of their humanity tangible in the theatre, he opens a window across national borders, beyond hostile authorities and detention facilities. He allows us to see the world where a person is forced to step into the inch of water on the floor of an orange dinghy and pray.

47 VANISHING POINT
Larissa FastHorse (2016)

'Then the oil and gas people cut more channels and sucked out everything that was under our little bit of remaining ground. Deflating the land so that it slowly drowned from its own weight.'

This one-act play by Larissa FastHorse deals with two cousins standing on their tribal land as it sinks below the waters of the Gulf of Mexico. Ilse de Jean Charles belongs to French-speaking residents of American Indian ancestry, who have lived for centuries in the Louisiana bayou, fifty miles south of New Orleans. The story is based on the tribe, the Biloxi-Chitimacha-Choctaw, who were declared the first official Climate Refugees in America in 2016. The federal government awarded them a settlement to relocate. Larissa FastHorse, who is an enrolled tribal member of the Sicangu Lakota Nation, has chosen to write about not just one tribe but the many people who have been affected by coastal erosion. There are an estimated two hundred million people in danger of being displaced by coastal erosion by 2050.

This play deals with a slow-moving disaster and the agonies of having to leave your tribal lands. Where can we go? How do we quantify what we are losing? The disaster is slow enough to prolong the dilemma but nonetheless shocking in its speed. Land the size of a football pitch is being lost every forty-five minutes. Since 1955, Ilse de Jean Charles has lost 98% of its mass. It was once wholly sustainable, so much so that the island's residents were unaware of the Great Depression because they could happily trade with each other, hunting and fishing from the natural bounty of the bayou. The combination of natural erosion, collapsing and compressing land from oil drilling and pipeline construction in the sea bed, and the loss of the barrier islands further out in the bay, have all combined to reduce this community to almost nothing. Only fifteen residents remain and half of their dwellings were destroyed by Hurricane Ida at the time of writing.

FastHorse has a natural resilience and an unbreakable tensile wit. She uses satire and seriousness to unsettle and provoke audiences, allowing them to land somewhere they might not have thought they could. She can embrace the entire magnitude and depth of a story and wrangle it into surprising and profound conclusions. She is a writer who sees the playwright as just one part of a vast network creating community bonds.

48 FURTHER THAN THE FURTHEST THING
Zinnie Harris (2000)

'Something terrible under the water, underneath where it is dark. Mill, I's heard something...'

A modern classic, Zinnie Harris's compassionate play captures a unique community in a persuasive and archaic vernacular, their sense of place strongly evoked in her beautiful language.

The play tells the story of the Pacific Island Tristan da Cunha, the most remotely inhabited place on Earth. Its population of around two hundred and fifty was largely a subsistence community formed of the descendants of nineteenth-century sailors. The First World War entirely bypassed the island; no ship had docked for a decade. There is still no airstrip. The journey takes six to eight days from South Africa, the crossing entirely dependent on the weather. In 1961 the eruption of the island's volcano caused a mass evacuation. All of the island's inhabitants, human and animal, took to the seas in their open fishing boats before being rescued by larger vessels.

The opening act of the play takes place in the days before the eruption. Mr Hansen, an entrepreneur, has arrived on the island with the purpose of setting up a glass jar factory. The island community have mixed feelings about this proposition. The minister, Bill, has been swimming in the volcano's caldera lake and is convinced he senses something bad about to happen. His nephew Francis, however, believes modernisation must be welcomed. The picture painted of the factory, with its conveyer belts and containers, everything quantified and accounted for, is a perfect metaphor for this last outpost of nonconformity to be brought into capitalism's embrace. The proposal is voted down in an island meeting but before the dust can settle, the volcano erupts and the islanders are forced to evacuate.

The second act takes place in Southampton, where the islanders have been rehoused by the British and given jobs in Mr Hansen's UK factory. He lies and tells them the island has been rendered uninhabitable by the explosion and they are unable to return. A darker truth slowly emerges, that the British have actually taken the opportunity to use the island as a test site for weapons. Distraught, the community must decide whether to face a different future here or fight for their right to return.

The play is elemental in its language and heartbeat, burning with the pain of a community ripped apart by both money and the Earth's forces. There is nothing romantic in Zinnie Harris's description of island life (based on her own grandparents' experience of living there). She is unflinchingly honest about the hardship and social deprivations of life further away than the furthest thing. Nevertheless, the sense of community and belonging, to each other and to place, is heartbreaking to witness, and devastating to see being destroyed. This is what happens when the outside world comes calling for economic and territorial gain.

49 PARADISE
Kae Tempest (2021)

'These days, it's just a vacant place –
An island rich with human poor
The world dumps all its plastic waste
And takes a couple hundred more
Fair exchange, I suppose
We're all just rubbish on the shore.'

Kae Tempest's play is a revelatory new version of Sophocles' *Philoctetes*, the story of an archer who fought in the Trojan War only to be betrayed and left to die on a desert island. We follow Neoptolemus, a soldier on a mission to persuade Philoctetes to return to the front line. The prophesy that only Philoctetes' peerless skill with a bow and arrow can win the war has convinced Odysseus that this archer must come back to the fray – at any cost.

The play is set on the island where Philoctetes is ten years into exile. His cave is on one coast and a refugee camp is established on the other. In this camp live a female chorus of climate and conflict migrants. This chorus are front and centre of *Paradise*. They share their worldviews, experiences and lifestyles, and emerge as the ethical heart of the play. This was once a busy place, full of life, love and activity, but now years of resource wars and extreme weather have left it a dumping ground for unrecyclable waste and unwanted people. A cacophony of sirens and alarms disturb the action.

These displaced women have created their own vibrant community. They forage for food, listen and sing stories, and wryly help each other make sense of the world. They possess an innate sense of justice and a global and timeless wisdom. They observe the arrival of Odysseus and Neoptolemus and watch the events that unfold with a sharp humour about the ludicrousness of conflict and the risible male egos who instigate such wars. They are a compelling theatrical force. Alongside Philoctetes' possible return to his warrior state, their own experiences show what it is like to exist on the periphery of a society ravaged by unbridled conflict. How do you keep your sense of self and identity? The play questions if in these circumstances we can ever – or would ever want to – return to the life we once knew. The conspicuous courage and value of these women transcends their 'unwanted' status. Their piercing reading of Philoctetes' dilemma becomes our understanding too.

The visual and aesthetic freedoms of the play match the heightened emotional currency. This is a world of flatscreen TVs and ancient prophesies; traditional medicine and migrant documentation. It is a charged and visceral drama with modern comedy cutting into moments of intense Greek tragedy. The play offers us an ebullience of searing poetry with physically bold, indelible set pieces of violence and action.

Paradise precisely and deliberately transforms a traditionally anonymous chorus into a stage full of distinct women who step towards us with their individual stories of survival. Tempest's trio of male characters (played by women in the original production) fret as much as they strut, indulging in rituals of pride and ego. Their loyalty to repetitive cycles of conflict is held up to the light, and we witness its volatile pointlessness. Sophocles would have understood. He set his story of Philoctetes on Lemnos, a Greek island in the North Aegean Sea, which today is used as a landing point for refugees from North Africa and the Middle East. Endless resource wars have driven them through this place. In this insightful new version, hope and strength are to be found in the resolute creation of new community and in the necessary fellowship of each other.

50 CANNIBALS
Rory Mullarkey (2011)

'The world's not overflowing, you know. There isn't enough, woman. It's dog eat dog, and living means taking from others.'

Cannibals is a play of mud and blood, animal skinning, and escape, a brutal, visceral study of human trafficking and the cost of reinventing ourselves time and time again. Rory Mullarkey asks us to face one very uncomfortable question: at what point might our humanity run out?

Farm life on the eastern edge of post-Soviet Europe is hard. The already-struggling community are scavenging the animal world for meat when their world tips into conflict. Soldiers arrive. Lizaveta's husband is shot. In a split second, her whole world turns. She makes an instinctive decision to run, shape-shifting to stay alive, trying to pass unnoticed, desperate to find safety. Instead, she finds herself in the hands of a human trafficker, sold like a commodity down the chain of traffickers to England. In Manchester she is now 'Tim's' wife'. What appeared to be a history play from the early twentieth century, suddenly and shockingly reveals itself as a story of now, a dazzlingly disorientating theatrical coup that pulls the rug from beneath our feet.

There are many hidden climate victims, like Lizaveta, forced to abandon their lives amidst sporadic fighting that is never officially defined as war. Human trafficking is a particularly heinous phenomenon that preys on those already disorientated. Lizaveta journeys from eastern to western Europe, household to household, village to city, endless, circuitous landscapes in which she can only adapt to survive. The neon lights of her destination city are a world away from the simplicity and silence of her homestead and animals. How much time has passed? A week? A year? *Cannibals* is a vivid, unsettling smack in the face. Mullarkey choreographs its extreme imagery, non-stop threat and brutal language to share the extremis and loneliness of Lizaveta's unsought life.

Cannibals can be seen as a companion play and unintended sequel to *The Grain Store*, inspired by a black-and-white photo of victims of the Holodomor forced to eat human flesh or starve.

THE SUPPLIANT WOMEN 51
Aeschylus (420BC/2016)

In a new version by David Greig

*'You guided our boat here, you gave us beach breezes
And now, Zeus, we need your protection again.'*

This play does everything we need a play to do right now. It gathers the community together, blurring the line between audience and performer, allowing everyone to experience the terror of refugees fleeing their homes. Crucially it places those experiences within the complex web of political considerations that make accepting asylum seekers into your land so fraught. The fact that this play exists at all is exhilarating. For 2,500 years it lay in fragments and scattered reports of its existence, until David Greig reassembled this new version.

Fifty women escape their homes, leaving everything behind. They board a boat from North Africa to cross the Mediterranean to Greece. They could so easily be the women of Afghanistan fleeing forced marriage and cultural privation in their homeland. They believe they can find protection and assistance in this new country; this after all is the message that wafts across the water: that the Greeks have a form of justice and a democracy that obligates them to act morally. Optimism propels every stroke of the oars. They land on the beach to seek asylum.

Aeschylus's *The Suppliant Women* is believed to be one of the world's oldest plays. Greig's new version allows the same agonies and expediencies to speak to us through the ages. Its relevance is shocking. These women are fighting for their lives. The King thinks that if he accepts these women it will provoke war with Egypt.

There could have been many more Greek classics in this collection, *Antigone*, *Oedipus Rex* and *Prometheus Bound* to name just three. That ancient civilisation's position in the Mediterranean, their self-inflicted ecological degradation, expansionist mindset, prolonged wars, subjugation of slaves and women, and their political, cultural and philosophical sophistication, allowed them to rehearse and reflect on many of the struggles we now face. The difference is that they were dealing with a discreet part of the globe. Now the lessons they learnt are applicable globally.

The power of *The Suppliant Women* is unleashed by Greig's decision to build this play in the heart of the community. The majority of performers on stage are the chorus. They drive the action through their songs, dance, rhythmic text and appeals to clemency. Each production is encouraged to find the performers for the chorus in the city or town where the production is being made. The accessibility of this process allows each production to shape-shift to fit the community. It draws the concerns of each locale, poultice-like, into the show, and this gives the truth of the story an irresistible power. This is the inclusive theatre of the future.

We have built a world where a care worker and a politician do not know each other, do not have any common ground on which to speak of their lives. *The Suppliant Women* is key to revitalising our theatre culture, building a joined-up, multilayered, enriching, supportive culture that embraces and empowers everyone, and relishes the dynamism of diversity.

52 FAMINE
Tom Murphy (1968)

'I was born here. I'll die here. I'll rot here.'

Through twelve extraordinary scenes, playwright Tom Murphy offers snapshots of different members of the same community at a moment of shared crisis. All of them are offered a choice, to stay or to leave. All of them are destroyed as much by the act of decision-making itself as the outcome of their choice.

Set in the village of Glenconnor during the Irish famine of 1846–47, the play follows several different households as they face the bleakest of dilemmas. John Connor is head of his family and de facto leader of the village. His ancestry in this place goes back generations. Yet everything he sees around him is crumbling. The second crop of the year has followed the first and failed. Poverty, starvation and death seems to be the only future left. Landlords, deaf to any circumstantial pleading, demand the rent must be paid. Three choices are on the table: to accept a political offer of free emigration to Canada; to sell oats that are only just harvestable to the English, who will ship them to the mainland; or to hold firm, eke out an existence, and pray.

This is dark territory. *Famine* faces head-on the bleak rituals of death, wakes and mourning that characterised the period known as the Great Hunger. The choices on offer are not of equal consequence. Enforced emigration has repercussions down the generations. Selling all they have is a destructive act of communal sacrifice. And holding out is to condemn yourself and your dependents to a slow and painful death. The most devastating scenes of the play show people begging their loved ones for a merciful release from living.

It is sobering to think these same rituals of wakes, grief, murder and mercy-killing are repeated elsewhere in the globe even today. In the Horn of Africa, as people face yet another year of drought and poor harvest, they find themselves in the same bleak world Murphy describes.

Tom Murphy's deeply humane characters scavenge and rage, suffer with nobility and pain, falter and waver. This is an insight into human suffering on both an intimate and Shakespearean scale. It is written with compassion and courage. All any family can do is choose what is right for them and their God. *Famine* is a landmark play of Irish theatre that has become global in its resonance. Worryingly, it will never date.

PART 7
RESPONSIBILITY

Responsibility

'It's hard to keep apocalypse consistently in mind, especially if you want to get out of bed in the morning.'

Zadie Smith perfectly articulates how exhausting and impossible it is to live in a state of constant vigilance. We all know we need to do our bit, but once you start taking responsibility it can feel like a full-time job. And then when we see someone else not shouldering any of the burden we can be tempted to stop bothering ourselves.

Don't use plastic cutlery. Easy. Only buy line-caught fish from inshore waters, fine. But who is going to dredge the ocean to remove the plastic fishing nets that are the real problem? Why was there 30kg of netting in the stomach of that dead whale? We were told plastic straws were the problem. In fact, aren't we being sold a lie by our governments? They emotionally blackmail individual citizens to take responsibility, but isn't it actually big business that causes the damage? Why is it all on us?

The instinct to blame someone else is gorgeously, momentarily comforting. Blaming a neighbour for not recycling, or some faceless corporate boss for exploiting everyone, or CONSUMERISM generally; it feels good to outsource the problem – and to outsource the effort to fix it. So how do we stick diligently to our personal task list, whilst keeping our constructive criticism for the more egregious offenders?

Developed countries and the industries that bolster their economies have historic responsibility for the vast majority of emissions and waste. The rapid increases in carbon output from emerging economies is still negligible in comparison. Rich countries can now speed up the transition to cleaner fuels and technologies, outpacing poorer countries and putting pressure on them to catch up. But poorer countries cannot do this overnight. They do not have the means or the infrastructure that developed countries take for granted. How do poorer countries invent overnight their green technologies from a standing start? How do governments convert abstract historical obligations into concrete cross-border actions, at pace? What is a fair share of responsibility? And where are the social thinkers who can save us from the tangled Virtue Olympics that beset these conversations?

The plays in this chapter all look at what taking individual responsibility really means. From Duncan Macmillan's exquisite *Lungs*, which follows a couple wondering whether to start a family, to Lucy Kirkwood's heartbreaking *The Children*, asking an older generation to make the ultimate sacrifice, these plays ask what it takes to personally step up. Should an individual be required to make amends for past mistakes? How do we dutifully map out our futures in light of knowing what we know? How much responsibility can one person actually bear?

LUNGS 53
Duncan Macmillan (2011)

'I could fly to New York and back every day for seven years and still not leave a carbon footprint as big as if I have a child.'

To breed or not to breed, that is the question. This brilliant two-hander, written to be performed on a bare stage with minimal design, is a low-carbon play about having a big carbon problem.

Short of building your own jet engine and flying it around the world once a week, having children is pretty much the most environmentally damaging thing a person can do. An individual is estimated to create about 10,000 tons of CO_2 over the course of their lifetime. Framed in this way, bringing another human into the world is pretty irresponsible.

The couple in the play go back and forth over this dilemma, frighteningly relatable overthinkers who might be deflecting from their inability to commit entirely to each other by way of a climate emergency. In funny, edgy, smart encounters, the two nameless characters riff off each other's anxiety. We watch them prepare to take the next step in their relationship, only for it to brutally fall apart. *Lungs* is unflinching about the realities of love, pregnancy and heartbreak, all the while circling around the endless dilemma of how far we should take the idea of personal responsibility. When does that thought turn into personal sacrifice?

Eight billion people currently live on this planet. It took two million years for the world's human population to number one billion and only took another two hundred years to race to seven billion. This staggering increase is the reason behind all the other reasons; every problem can be traced back to population. There are simply, and suddenly, more people on Earth than the planet can sustain. Attempts to restrict population growth have opened very large cans of legal, ethical, social and religious worms. Dissecting this thorny issue at the same time as dissecting their relationship, the characters embrace their own fragility in order to find a liveable truth.

A love story with a twist, Duncan Macmillan's play is an actor's dream, a masterpiece of graceful, deft playwriting. Perhaps the first step on the road to facing an uncertain future is to find someone to share it with.

54 CROWN PRINCE
John Godber (2007)

'Oh don't you start. It's all a load of my arse all that, there's been heatwaves and ice ages on and off for bloody years. I've seen it on telly. It's just the latest fad. It'll be sommat else before you know it.'

John Godber's classic triumvirate of sport, comedy and politics is used to stonkingly good effect in this funny play about a bowls club in denial. Their perfect green could never be compromised by climate change. Nope. Surely not. Nah.

Jack was once a champion crown bowler and doesn't let anyone forget it. But he's not won a match in years and is regularly beaten by the club's smooth-talking heart-throb, Ronnie. Caroline, the club's resident top lady bowler, teases both men. Ted, Jack's friend, tries to keep it civil and everyone on an even keel. Jack's teenage granddaughter Faye comes to pick him up after a match. Faye is a fired-up climate activist. She blindsides the entire club with her ecological doom. Affectionate but adamant they will be okay, the club carry on bowling.

There are significant time-jumps between the following three acts as we see these likeable characters age from sixty-one to eighty-one. Over those twenty years, the climate worsens and the weather becomes more volatile, until, in the final act, much of the surrounding area is under water.

Hull is a familiar city setting for John Godber's body of work, but in *Crown Prince* he tells a less well-known side to Hull's locale – that the city and the East Yorkshire landscape around it are extremely flat and low-lying. Along with Amsterdam and New York, it is dangerously vulnerable to rising sea levels.

The bowlers are protective of their rules, their green and (despite their competitiveness) each other. As the waters begin to rise, their little club on its small hill becomes a place of refuge, a beacon of hope for them all. The play is packed with great gags, cultural and political references, and some very funny set pieces (such as when the group try cannabis to ease their arthritis and they get the proper giggles). The slow-burn time frame cleverly reveals how we knew about all this ages ago and did absolutely nothing.

Through each act we watch the bowlers, already of a certain age, get older. Ailments and marriage problems are discussed as well as their slowly changing worldview. Faye's predictions from fifteen years ago begin to land as they come true. Ronnie, now eighty-one, turns up carrying a paddle in to the clubhouse along with his walking stick. He has arrived in a boat. Jack, now a cycling great-grandfather, is a full-time climate convert and is going on a march with Faye tomorrow (after a game of bowls, obviously). As Caroline drily remarks in the final act of the play: '*People only understand politics when it's happening to them.*'

THE FORCINGS 55
Kevin Artigue (2021)

'Seventeen protestors, all young, graduate students, were ordered onto a bus and driven away from the protest site... this was four months ago... they're called the seventeen.'

A family get-together in Mexico turns into an eco-political thriller with a touch of magic realism, when a stranger arrives claiming knowledge of the family's missing daughter.

Sofia went missing a year ago and her family don't know if she is dead or alive. Her parents and two sisters come together to mark the occasion, each of them at a different stage of grief and denial. One sister, Pilar, is newly pregnant. She is concerned about the ethics of bringing a child into the world. She is married to Kev, a PhD ornithologist. Birds keep dying on the beach and he doesn't know why. The other sister, Rosie, works in corporate responsibility with Exxon Oil, the company her father, Ernie, recently retired from. He now spends time following up leads on his missing daughter's whereabouts, while his wife chooses religion as her way of maintaining hope.

The history of fierce battles between activists and oil companies casts a long shadow over this play. Since environmental organisations first began direct action, the two have clashed, often fatally. Global Witness reports the number of climate activists killed has increased dramatically in recent years. There were 277 murders in 2020, including Fikile

Ntshangase, a South African grandmother shot for leading a campaign against a coalmine, and Óscar Eyraud Adams, killed for advocating for his community's right to fresh water in Mexico.

Activists try repeatedly to prevent new pipelines from being built, or to disable existing infrastructure. Historically this was done in person, more recently it has moved online. In response, oil companies use heavy-handed tactics and illicit methods to defend their property. Corporate and personal responsibility collide in *The Forcings* as Ernie is made to confront the news that his missing daughter was protesting against one of his own pipelines.

The play allows us to watch this well-established middle-class family, whose lives are built on the largesse of the oil company, come to terms with the fact that this same beneficent organisation is responsible for Sofia's abduction. A stranger, Oscar, appears. He loved Sofia and saw her murdered by a crime cartel employed by Exxon. He confronts Ernie, forcing him to admit he personally ordered the killings. He never imagined that his own daughter would be among the activists. A tree cracks in a thunderstorm and falls, scattering branches across the garden. The sea, not far from the house, begins to retreat as it does before a tsunami. This family, picking over the pieces of their shattered life, feels suddenly deeply vulnerable. The visceral terror of battening down the hatches and trying to protect your own family is horribly convincing. As the winds pick up, we know that disaster is on its way. This is an emotionally charged play about how the climate crisis can be located within a single family.

56 THE WEATHER
Clare Pollard (2004)

'It's natural, it's cycles, it's unproven, it's cows farting, it's China; it's probably China.'

Clare Pollard is a poet and playwright not faint-hearted about the climate crisis. *The Weather* is a biting black comedy which asks if one generation's recklessness has curtailed the next generation's potential.

Ellie is a teenager crippled by anxiety. She is trying to get on with her life and even start dating, but at every turn her parents conspire to hold her back. Her mum Gail is a narcissist whose moods swing wildly between euphoria and depression, while dad Bob is all about money.

Every scene takes place in extreme weather. It can be snowing in July and sweltering in February, one scene stormy, the next merciless sunshine. It is theatrically disorientating and unsettling. The timeline of the play is season-less, the weather as erratic as the character's behaviour. Gail and Bob keep on spending, debt keeps accumulating, the mountain of stuff keeps growing. With all seasons gone, the normal rhythms of the year absent, Ellie's parents too become increasingly unreliable.

The Baby Boomer generation, which describes people born between the end of the Second World War and the early 1960s, supposedly had the best the twentieth century could offer – peace, prosperity, education, advantageous economics and culture. Contrast that with their millennial children, growing up in an age of cyber warfare and terrorism, cut loose from the property ladder, education that must be paid for, and a cultural atomisation that splinters community. It is a perfect recipe for intergenerational conflict with the climate crisis as the chosen battlefield. The period known as the Great Acceleration are the years of post-war activity, furious rebuilding, rapid industrialisation and a steep upward curve in consumer behaviour. It is a prime source of climate rage and a wellspring of activists' inspiration. But it is also the era that led to universal social benefits, medical breakthroughs, and technological innovations that most people would struggle to live without.

However, nuance is not the vocabulary of *The Weather*. This story takes a turn towards horror as the family's buried trauma unleashes a poltergeist. All of the voraciously acquired consumer goods fly around the house. The deep-rooted guilt driving the family's dysfunction is revealed to be domestic abuse, with Ellie bearing the brunt of her parents' rage. This dark metaphor captures the generational discontent and the corrosive effect of the Boomers' incessant consumerism. Ellie and the house fight back, Ellie, now newly empowered, knife in hand, stands ready to take responsibility and wreak her vengeance.

57 IF THERE IS I HAVEN'T FOUND IT YET
Nick Payne (2009)

'I'm pissed off that you think I'm some kind of ecological heathen because I bought a tiny tin of pineapple for the first time in six months.'

Nick Payne's sharp and funny play is a compassionate dive into the responsibility dilemma. Do we act at the micro level (domestic and personal) while keeping one eye on the macro (global and massive), or do we campaign loudly against the macro whilst letting our daily attention to the micro slip? This is the story of a family trying to save everything but themselves.

Fifteen-year-old Anna is being bullied. She's overweight. Her mum Fiona is embarrassment personified, she's even stepping in to help on the school musical of *The War of the Worlds*, and her dad George is too busy writing a manifesto response to the failure of the Kyoto Protocol to be a parent. George's book calculates the carbon footprint of *everything*: wine, avocados, pets, sheet music, sticking plasters. Anna headbutts the bully-in-chief at school and gets suspended. Terry, an unlikely saviour in the form of George's long-lost weed-smoking brother, steps in to take a left-field interest in Anna's life and education. He teaches her kung fu and takes her to museums. Meanwhile, George's constant critique of the carbon footprint of the weekly shop is driving wife Fiona away. They need to spend more time together. Perhaps a holiday? But can George be persuaded to get on a plane?

According to climate researcher and author Mike Berners-Lee, an average paperback book, printed like this one, and assumed to sell out its limited print run (tell your friends!), contributes about 400g of CO_2e to the Earth's atmosphere. Our publisher cares about these things, so you are holding an above-average impact-light slice of publishing in terms of its carbon footprint. (You'll find some information at the end with details about how this book was printed.) But can we calculate the footprint of everything we interact with? The minutiae of such an approach can be as overwhelming as the vastness of the problem. *If There*

Is I Haven't Found It Yet mines this rich seam of equivocation. If we angst too much about the macro, do we risk not seeing the concrete difference we can all make in our homes? And if we only focus on the tins of pineapple, do we lose perspective on the evil bastards offshoring profits from fossil fuel? This is a fantastically entertaining play trying to figure out what, or who, or where, or how we really need to concentrate our efforts, whilst satisfyingly never questioning why.

58 WHEN THE RAIN STOPS FALLING Andrew Bovell (2008)

'Their heads are bent against the relentless weather and against their relentless lives.'

This is the story of five generations of one family stretching from 1975 to 2039. The play, set between England and Australia, begins in the future before moving backwards, switching time frames elegantly. Gradually, its Möbius strip structure falls into place, culminating in a moving story of forgiveness, humanity and hope.

In Andrew Bovell's 2039 the world is coming to an end. The rain in Alice Springs is unremitting, the flooded streets making it impossible for anyone to enter or leave the town. A fish lands from the sky at Gabriel York's feet as he prepares to meet his long-lost son. Back in London in the 1980s, Gabriel's grandmother Elizabeth struggles to communicate with her son, Gabriel's father. Rewind further to the younger Elizabeth unable to speak to her husband. Fast-forward to Australia to meet Gabriel's mother, steeped in regret for the life choices she has made. The play skilfully dances between the story of each generation as they come to terms with how the life they have lived fails to match the one of their dreams.

This is a domestic play with a huge heartbeat, inventively paced, cleverly constructed, and packed with emotional integrity and depth. Across four generations of the York family, people constantly discuss the weather and console themselves that things could be worse, '*still, there are people drowning in Bangladesh'*. This backdrop of cataclysmic weather events provides a platform for Bovell to interrogate what the world can survive, and what we can survive of our own pasts.

The deterministic nature of time is the concept that all events are a result of previous actions. It's the opposite of randomness. In climate circles, determinism suggests taking responsibility for any and all future outcomes, rather than passively assuming the Earth will move on to another phase in its long cycle of natural environmental change. The fish that falls from the sky in this play is a symbol of something that appears utterly random yet actually has a reasonable, if uncomfortable, explanation. Alice Springs is a town nearly a thousand miles from the ocean in any direction, in the heartland of red-earth Australia, yet the fish still smells of the ocean. When tornadoes cross large bodies of water they suck up sea life into their vortex, blowing them into clouds from where they can be transported huge distances. With increasing turbulence in global weather patterns, this phenomena of 'animal rain' is predicted to become a more common occurrence.

Likewise, the events of Gabriel York's life appear completely haphazard, brought about by coincidence and chance, until each step of his familial story becomes clear to him, showing how the actions of one generation cannot fail to have an impact on the next.

As the play reveals why he ended up in Australia and how he became the man he is, the audience learn the multiple ways in which the past shapes the future, environmentally and emotionally. Beautifully written, with lyricism and subtlety, the way in which the characters finally face up to the truth of their troubled pasts makes for a wise and hopeful ending.

59 EARTHQUAKES IN LONDON
Mike Bartlett (2010)

'Er sorry to interrupt you but I've had enough of the environment, hear about it all the fucking time, I only did it for my sister and she didn't even turn up.'

The ultimate climate-crisis party play, *Earthquakes in London* is a five-act spectacular. Playwright Mike Bartlett picks a fight with *epic* and doesn't back down.

As Robert, a climate scientist, knows, London experiences frequent earthquakes. Most register somewhere between 2 and 3.5 on the Richter scale, too small to cause much alarm. The largest one in recent years hit a 4 (the number at which geologists start to take notice). It happened at a depth of about 10km and just caused a few windows to smash. For Robert, these earthquakes are only going to get more frequent (the British Geological Survey agree with this position). Cassandra-like, Robert predicts a full apocalypse any minute. Not such a fun dad for his three daughters, from whom he is estranged.

The eldest, Sarah, is an environment minister in a coalition government trying to stop airport expansion. The second, Freya, is a suicidal teacher, about to give birth and terrified of the future her unborn child will inherit. And the youngest, Jasmine, is full of rebellion, a hedonistic student who has been expelled from university. She gets blackmailed by a lover who wants to influence her big sister's ministerial decisions. The play follows all three sisters as they navigate family tensions and impending doom.

This play goes big. It delights in its own excess. Characters rampage through a city sequence of time-jumps as fractured as the family and the planet. The story gathers scenes from the past (1968), the future (2525), and everything in between. Songs, dance, drinking and clubbing blur reality, hallucinations occur, stories shape-shift, futuristic happenings which are both real and illusory. The play is at times beautifully intimate and domestic and at others wildly kaleidoscopic and public, tiny family squabbles play out on a global canvas.

Robert slowly comes to terms with his own guilt about decisions he made early in his career, when he diluted research findings for a fat cheque from the aviation industry. His defiance and then regret seeps into the bones of his family and his daughters. Each in their own way, they define themselves against his past. The need to make amends is an inheritance we may all have to live with. So let's dance.

60 THE CHILDREN
Lucy Kirkwood (2017)

'Yes so then I wanted to call Robin so I walked, I ran down to the beach, because the reception – and that's when I saw the tide had gone out. I mean it wasn't miles but it looked like miles, and then I saw the wave, only it didn't look like a wave, it looked like the sea was boiling milk and it just kept boiling and boiling and boiling.'

The Children is set in a small cottage on the east coast of England in the near future, where a recent disaster at the nuclear power station has devastated the area. Robin and Hazel want to continue living in their house despite knowing the dangers of nuclear fallout. They are retired from their jobs as scientists at the plant they helped to build. Electricity and water are rationed and they keep a Geiger counter to check for signs of radiation. Hazel is determined to preserve some semblance of normality and lives the healthiest life she possibly can. Robin disappears daily to feed cows who miraculously survived. He brings home salad that Hazel checks before devouring. Rose, a former colleague, whom they haven't seen for thirty-eight years, turns up at their door. She seems intent on disrupting their precarious but ordered existence.

This is a genuinely disturbing play. Lucy Kirkwood's three-hander pulls no punches. It asks the central question of this global crisis, unadorned. Will we take responsibility for our past actions? The scenes have the visceral tang of life lived in a destroyed landscape. Gradually we realise why Rose has come back. She wants the young scientists and physicists who are currently in the leaking nuclear plant, working to make it safe, to be relieved by those who are over sixty-five. Having built it, they should be the ones to secure it. She needs twenty people. She has eighteen. She has government approval. Robin and Hazel now know why she is there.

Hazel goes mad. She has four children. She is a good mother. Why can she not be allowed to see her grandchildren grow up? She vents her anger about Robin and Rose's past affair. Robin admits that he has been returning to the farm every day not to feed the cows, as Hazel was willing to believe, but to bury them. He coughs up blood, setting the Geiger counter off noisily. He is happy to agree to Rose's request. Is Hazel?

The bonds between these three people make their long-standing obligations towards each other very present in the room. And this in turn allows them to face their obligations towards future generations. This is a tough play, brilliantly effective.

PART 8
COST

Cost

We use economic growth as the barometer of human happiness. We look to the FTSE 100 and Wall Street to see how stable and prosperous we are. But maybe we need other measures, other ratios, other bar charts to understand what is really happening. The idea of living harmoniously with nature is increasingly at odds with relentless growth. How can we see this, how can we know this, how can we make this legible in our thinking long enough to act on it?

Plays are a way of making alternative worldviews vivid and lasting. How do we open up our ethical and imaginative thinking, make our thinking supple enough and resilient enough to find the breakthroughs we need? How can we possibly live sustainably and lightly on the planet whilst the invisible hand of the market urges us to ever-higher financial targets and higher standards of living, whatever they are. What are the other planetary narratives that needs to be loud in our communities, articulate in our conversations and lucid in our daily interactions? We are caught up in a race with no end point and with constantly shifting goalposts. We are trapped in a politics that is diminishing the arts at exactly the moment we need the empathy industries to flourish. The vocabulary of growth is not the vocabulary of the next few years. How do we get our minds and mouths around some new language?

The degrowth movement advocates social and environmental wellbeing as a better and more useful marker of genuine prosperity rather than GDP. It suggests the very idea of 'sustainable development' is an oxymoron, designed to disguise continuing economic growth window-dressed as a green future. Instead, the idea of degrowth backs a circular (rather than linear) economy. It is built on recycling and reusing, localism and commons, community and non-profit. It is such a shocking set of new thoughts that it is attacked for being anti-business by business, and for being anti-workers by unions. This idea, like so many other proposals to save the planet and find alternative ways of thinking and living, finds itself in the crossfire. This is the kind of evolution in our thinking that cannot be left to governments and the coming STEM generation. We need arguments forged in the crucible of theatre. For and against, outraged and inspired, revolutionary and lasting. Degrowth might not be the whole answer but it is just one of the useful frameworks through which to interrogate how we balance environmental protection with people's economic, physical and emotional wellbeing.

The plays in this chapter look at the cost of unchecked progress, on both humans and the environment. They offer historical perspectives on fights we thought we had won, on mistakes we thought we had legislated away, and on the test of characters under extreme societal pressures. Erika Dickerson-Despenza's astonishing *cullud wattah* is a look at the ongoing Flint water crisis while Karel Čapek's troubling *R.U.R.* considers the implications of an automated future.

61 cullud wattah
Erika Dickerson-Despenza (2020)

'white folx'll take this house n flip it for a million once this whole lead thing is ovah'

Cullud wattah is a poetic, visually arresting play that packs a huge emotional punch. It follows the lives of a family of women caught up in the ongoing water crisis in Flint, Michigan, a majority black US city. Erika Dickerson-Despenza is a playwright with the future in her sights. Her play is fearless about how engaged we are invited to be. Audiences are given a bottle of dated, contaminated water to add to the stage as they enter. Maximum empathy is required and resistance is futile. You cannot help but be swept up in this rhythmically compelling story. You cannot help but be a pool of liquid on the floor at the end.

At the time of writing, it's been 2,755 days since Flint has had clean water. The city administration saved money by piping water from the Flint River rather than the water treatment plant, but the water was polluted, full of lead and bacteria. Now nothing comes out of the taps. The mains have been turned off. The house is full of bottled water. The walls of the Cooper house are covered in a tally of the days since the crisis began.

Marion and Ainee are adult sisters. Marion is a widow with two daughters, Plum and Reece, and her elder sister Ainee is single and pregnant. Big Ma completes the household. The family are riddled with poison, their rashes and ailments are numerous, their resilience astonishing. The sisters are torn about how to deal with this crisis. Marion is a third-generation worker at General Motors. She doesn't want to cause trouble or she will lose her job. Ainee wants to join a class action and take the city to court.

The degraded river is at the heart of this story, poisoned by the local car industry. No one thought to admit it until it showed up on the skin of young women, and as lead in the brains and blood of schoolchildren. Ecological racism is the subject of this play, the intersection of climate degradation and malevolent societal neglect. The climate emergency is sited within these women's bodies. When the poisoned water pollutes the amniotic fluid of the child in Ainee's womb, the metaphor of the play becomes unbearably concrete. The Greek idea of pollution being 'of the soul' is captured with astonishing power by *cullud wattah*. The purest water we know, the life-giving amniotic fluid, mingled with the most corrupted water on the planet. It is devastating.

62 DELUGE
Fiona Doyle (2014)

'You've started a trend. Mick Dwyer's building one now. And his neighbour, Con Sullivan, he's doing the same.'

Joe is building an ark. The rain shows no signs of relenting and it's the only practical thing he can do to save both family and animals. But when the farm spirals into even more debt, Joe goes missing, and a weather crisis turns into a manhunt. Fiona Doyle's crime drama is a psychological thriller that walks us towards the darkest impact that the worsening climate has on the lives of farmers. Please be aware as you read on this play explores suicidal feelings.

Set in rural Ireland, the action of the play is split between a police interview cell, where Joe's wife Kitty is being interrogated, and flashbacks to the events on the farm that led up to her arrest. At first she is reluctant to talk, but the promise of seeing her baby son Lorcan proves too much to resist. She begins to reveal a troubled story of debt and fear. We watch in horror as the family's circumstances become too much for Joe to bear. The farm's financial problems escalate as the weather prevents any chance of the animals getting good grazing or dry fodder. Creditors circle the farm like vultures. Joe's paranoia increases and he shuts himself away in the barn, hammering away at his ark. Friend and farmhand Flan tries to help Joe and Kitty make sense of their situation, but when Kitty falls pregnant, it all gets too much for Joe.

The number of farmers who die by suicide is amongst the highest in any occupational group. Every case is different and support groups point to financial pressures, rural isolation, legacy burden, physical and mental health issues, and a total loss of predictability in the clock of the seasons. Agrarian economies across the world, such as the countries of eastern Africa, central Asia and in particular India, report similar statistics and circumstances.

In the play, Joe believes his suicide will enable Kitty to claim life insurance and thereby save their beloved farm. However, the small print reveals otherwise. The fear that the relentless rain will make the farm unviable brings on Joe's darkest moment. This is a compassionate and disturbing play about the hidden cost of not listening to the very people who know first-hand that the Earth is in pain.

RADIUM GIRLS 63
D. W. Gregory (1999)

'I don't want compensation. I want court. I want them to look at me and explain how it's my fault I got sick working at their factory.'

Radium was a miracle discovery. The excitement was unbridled. What could possibly go wrong? D. W. Gregory's play is an achingly familiar cautionary tale beautifully done. We all recognise that the unregulated exploitation of new science is just too tempting to resist, but to see it unfold in the story of the American Radium Girls is deeply affecting.

The Radium Girls were the female factory workers in the 1910s and '20s whose exposure to radium gave them radiation poisoning. The play follows a familiar arc. A company discovers its manufacturing process is harmful but does nothing. Workers fall ill. Company commissions rogue research absolving responsibility. Workers claim compensation. Company filibusters. Legal process takes years. Company quietly fades away. Workers die. But in following the great excitement of the new discovery and watching it wreck people's lives, Gregory creates a new archetype. She burns through this simple story till we feel it on our skin.

A sacrifice zone is a place chosen to dump waste, in full knowledge of the environmental damage that will be done. These might be places where the recycling from rich nations is outsourced to poorer ones, or the locality of a chemical plant. Whatever the geographic circumstances, the local population are unlikely to have been consulted. It is always a higher authority that is doing the sacrificing.

People can also be sacrificed. This play shines a light on the workers on the front line of these industries. They are staff who take on trust that the materials they work with are safe. The play challenges our complicity in not asking questions. How much do we really know about the harmful effects on humans and environment of our consumer habits?

Radium Girls is a fast-paced ensemble piece built on a sequence of encounters woven together with news reports, archive material and verbatim testimony. Its many successful productions prove that we never tire of stories that give voice to the voiceless.

64 ONE FLEA SPARE
Naomi Wallace (1996)

'That's the curse of this plague. It's stopped all trade. There's not a merchant ship that's left the main port in months.'

Naomi Wallace's play focuses on the story of William and Darcy Snelgrave, a wealthy couple preparing to flee their London home during the great plague of 1665. Bunce, a sailor, and Morse, a street girl, break in to their boarded-up home and the four end up enduring quarantine together, a month where their social, sexual and moral boundaries are questioned and crossed. This viscerally achieved play asks uncompromising and testing questions of our sense of 'society'.

All cities retain some form of racial or social separation in the architecture of their streets. Communities are cut off from each other due to the historic allocation of desirable space and access to resources. Privileged space and imposed boundaries are being rattled daily, the start of a climate-change time bomb. *One Flea Spare* explores how people from different backgrounds negotiate each other's space when those boundaries are forced to break down. This reluctant fellowship must cooperate if they have any hope of surviving the threat outside their walls.

Green cities have much to offer the world. Eco-modernists point to cities outperforming rural communities in their ability to share resources and reduce emissions. Cities also tend to have lower birth rates and faster modernisation. A move to shrink the human footprint on the planet to urban centres only, allowing nature to flourish outside the city walls, has gained some traction (as well as much criticism) in recent years. Urbanists suggest we cut our losses, divorce our material needs from nature, make citadels of self-contained human ecosystems and preserve whatever wilderness is left. This decoupling activity promises solutions in GM foods, aquaculture, synthetic meat, precision agriculture and centralised planning. *One Flea Spare* questions whether such separatism from nature is wisdom or folly. And how does social inequality fit in to this vision?

Naomi Wallace's unsentimental and poetic take on the politics of death is timeless. The play wonders what entitlement to personal space really means – is it the body, the home, or the neighbourhood? What universally understood civic rules need to operate? As we all know, shutting the door does not keep the plague at bay.

65 THE CHILDREN OF THE SUN
Maxim Gorky (1905)

'Be still Pavel, you can see nothing: you look through a microscope.'

An outbreak of cholera is sweeping the city. Cosseted in his laboratory, Pavel Protasov beavers away, oblivious to the social crisis unfolding in the streets, and of the emotional crisis building in his household. *The Children of the Sun* dares to suggests that the Enlightenment is an illusion. The triumph of knowledge over ignorance might just be a fallacy: information does not always equal understanding.

Yelena, Protasov's wife, is neglected and bored. She flirts with his friend Vaughin, a self-obsessed artist. In contrast, the wealthy Melanie worships the ground Protasov walks on and wants to throw money at his research. He is as unaware of her attention as he is of his wife's infidelity. No one notices the domestic violence perpetrated on Advotia by her brute of her husband, lab assistant Yegor. Only Protasov's sister Lisa seems to sense something is going on, but her worries are dismissed as 'nerves' by the others. The characters indulge abstract ideas of beauty and intellect while the unrest in the city mounts. Eventually an armed mob break in to the house to attack the man whose experiments they believe are the source of the cholera epidemic.

The intellectual elite – 'the children of the sun' of the title – are excoriated by Gorky in this play for not paying attention to what life is like for the majority. We are no Gorky apologists (he supported Stalin before turning against him in later years) nor anti-scientists, but the blindness of the privileged is powerfully portrayed here. It chimes today with misinformation and the pandemic, but also with a fault line that will be with us for longer. There is increasing tension between those convinced that science alone can provide answers to the climate emergency, believing that ever-higher investment will be the answer, and those who know that social interventions and behavioural change also have to be made now. Are governments right to divert focus and funding in the name of progress, while all around them ordinary lives are blighted for want of a small portion of those funds? A modern version of this play would be timely.

66 R.U.R.
Karel Čapek (1920)

'Within the next ten years Rossum's Universal Robots will produce so much wheat, so much cloth, so much everything that things will no longer have any value.'

As the Fourth Industrial Revolution appears over the horizon, is it too late to ask if artificial intelligence is the answer? (The first industrial revolution was steam, the second electric, the third automation; the fourth is framed as artificial intelligence.) This visionary play by Karel Čapek chases down the fantasy that has long tempted us. If robots replace us as workers, will social emancipation and environmental renewal be ours? Or will it speed up exploitation of the Earth's resources and leave millions in poverty?

There are many pros and equally many cons. It may be the key to cleaner living and help to significantly reduce emissions. Perhaps automations will make production processes more efficient and allow us to wave goodbye to high-carbon job-related travel. Much manufacturing is already on this path. Having once employed thousands of people on a production line, now networked automatons do the same job.

The Rebound Effect is in the cons column. Automation drives the unit price of production down, so more people can afford to buy more things, leading to an increase in production and emissions rather than a reduction. More of anything leads to more unsustainable practices. This can be seen very visibly with food production – cheaper food means more eating.

R.U.R. was written in 1920, when mechanisation shifted the economy from artisan creation to mass production, but Čapek set it in 2000, at the moment he predicted automation would evolve into artificial intelligence. A character called Rossum, a biologist studying marine life, accidentally invents an organic material that creates robots. A factory is quickly set up and the robots put to work. These robots (a word Čapek famously invented in this play, from the Czech *robota* meaning 'forced labour') are made from living material, rather than the collection of highly processed minerals and data we think of today. They can think for themselves. Helena arrives on the island representing the League of Humanity who wish to emancipate the robots. But as the technology advances quicker than the politics, a new advanced 'breed' develops. The apex creation is a female robot named… Helena. The robots quickly outgrow their human owners, take over the factory, and exterminate the human race. Robot Helena becomes the new Eve for a race of cyborgs that will inherit the Earth.

The economics of automation was something John Maynard Keynes worried over in the same decade as Čapek's play. Why have we taken one hundred years to explore the implications?

67 THE FIELD
John B. Keane (1965)

'I watched this field for forty years and my father watched it for forty more. I know every rib of grass and every thistle and every whitethorn bush that bounds it.'

John B. Keane's dark parable has shifted its resonance since it was written. Where once we might have felt morally ambivalent about the outcome of the tale, our current sense of environmental justice may have tilted this play into new ethical territory. One man's love for a piece of land, and the lengths he will go to keep it his, still unsettles us, but in subtly different ways.

Bull McCabe, one of the most famous characters in Irish drama, grazes his cattle on a prime piece of meadowland. It is lush and verdant and slopes down towards the river. He has rented this land from the Butler family for decades, a happy, mutually beneficial arrangement. But with the death of Mr Butler, his widow is forced to sell. An auction is held. An out-of-towner with a lot of money comes to buy. He plans to concrete over the field to set up a warehouse for his building materials company. The plot's access to water is just what he needs – he can use the river to dispose of waste. Bull wants the land to keep grazing his herd, so he can pass the patch and the animals down to his son. He's saved up for a long time, on the unspoken understanding the land would pass to him. But his money won't stretch far enough in an open auction. Days of frustrating negotiations boil over in a midnight confrontation in the field, with fatal consequences.

The legitimacy of land ownership on an historically contested island is a profoundly emotive issue. Raising our eyes to the international resonance of this play, we see it applies to tenant farmers globally. They have no rights over the land they work, often for generations. Small farms are the backbone of rural communities. The focus on large-scale agribusiness and industrial development means they are often the first casualties in the race for efficiency. But the small-scale offers a much higher standard of soil health and animal welfare, as well as low-carbon, direct field-to-fork produce. Bull McCabe's cattle are what have made this field so special. It is their grazing and Bull's care over many years that have turned it from a raggedy patch of land into prime grassland, with rich soils and a symbiotic relationship with nature. It's no wonder that Bull loves this place.

Based on a true story, *The Field* shows how land is more than title deeds. It is identity. The village closes ranks around Bull and his son, cooperating with the investigation only as far as expediently necessary. Not because they think his actions are right, but because they understand that ultimately the land belongs to the community. It being cared for, and not destroyed, benefits everyone.

Rural life is no idyll. The play is far from romantic; it is pragmatic, even anti-pastoral. But it is not immoral. God is present and everyone wrestles with the rights and wrongs of what has happened. The whole community knows who killed the businessman. But no one will say, not for any money. This could be any small agrarian community anywhere in the world, willing to risk everything to hold on to their shared values. John B. Keane's hard-hewn parable is a classic play twisting in the storm-winds of history, showing us a new facet every time.

PART 9
EXTINCTION

Extinction

Natural history museums around the world are surprisingly short of dodo specimens. Very few people realised the poor bird was about to go extinct and so very few bothered to collect it. The small number of specimens that do survive are in pretty poor condition, studied for clues as to how the dodo might have moved, but unable to provide a complete narrative of its life. Some people believe it to be slimmer than it is traditionally shown in pictures, others think it must have been taller. But how did it raise its young? How soft were its feathers? What did it sound like? We will never know.

Biologists say we are currently living through the Earth's Sixth Mass Extinction (the Fifth was the dinosaurs). Thousands of plant and animal species are disappearing at accelerating rates and many more are classified as threatened or vulnerable. Recent losses include the Yangtze River dolphin and the northern white rhino. Habitat loss is the main cause of species extinction, as humans increasingly encroach into other creature's territories. Equally damaging is the amount of domestic and industrial waste that finds its way into fragile ecosystems.

Mourning individual species acknowledges their being and bears witness to their demise. But individuals are not islands; they belong to an interconnected web of life of which we humans are also part, our fate linked to the species with whom we share the planet. The plays in this chapter all look at the sobering facts of extinction in provocative and daring ways. From the animal-ghost of Lynn Nottage's exquisite *Mlima's Tale* to the selfie-hunting extinction tourists of Stephen Carleton's hilarious *The Turquoise Elephant*, these writers challenge us to notice exactly what we are losing. What does it mean to be the last of your kind?

The one single shift in our human mindset that would accelerate our response to the climate emergency is the acceptance that we are connected to the web of life. We understand connectivity in the virtual world, why do we struggle to understand it in the corporeal? The notion that we are supreme and separate beings is the cornerstone of our reluctance to change. We are not separate. When species extinction happens, we are on the list. Theatre is where mindsets can shift. The connection that actors feel inside the microclimate of a company on stage is something we recognise and value. The connection that audiences feel when the air in the theatre is charged with a shared understanding is how we listen deeply. We are experts in this. There is real work theatre-makers can do in this crisis.

MLIMA'S TALE 68
Lynn Nottage (2018)

'Know I stay away from you and the children, because I'm protecting you. My distance is my weapon.
I'm a shadow warrior all around you, listening to the sounds of the night. I hear everything. I hear you. I hear you.'

Mlima's Tale is an unflinching and deeply affecting fable that is as beautiful in its performance as it is troubling in its politics. Mlima is an elephant. Not just any elephant but a venerated big-tusked bull elephant, one of the last of his kind. Mlima means 'mountain' in Swahili, and Mlima is an important national symbol, loved throughout Kenya. Nottage's precise and monumental play follows Mlima's spirit from his murder on the open plains of the Kenyan game preserve to the pristine show cabinet of a Chinese penthouse, where his carved tusks now boast of high status, and a craven disdain for the ivory ban.

Mlima is protected, with an army of wardens patrolling the Kenyan savannah. He is instinctively staying away from his family as he knows he is being stalked. He is fatally injured, left in the dust to slowly die, and then before his last breath the impatient poachers axe his tusks brutally from his head. Mlima's tusks then pass through a sequence of various corrupt officials, custom officers, high-level diplomats, naive ship's captains, and a master carver who 'believes' these tusks do not violate the ivory ban. At every step Mlima is still spiritually alive, still suffering. Tusks from many herds find themselves in the ship's hold, spiritually aware that they are being shipped far from their ancestral homes. We feel every moment of their pain. The commoditisation of their lives is soul-aching, and carries resonances of the Middle Passage, the inhumane sea journey that took slaves from West Africa to the Americas.

The play demands a highly physical central performance to capture the spirit of Mlima and communicate the rupture from life, place and community. All of the human characters are meticulously drawn. They are the links in the corrupt chain, complicit in deed, deceptive in word, or self-delusional in silence. The sequence borrows effectively from Schnitzler's *La Ronde*, and at every transaction a mark of Mlima's spirit is left on anyone who has laid hands on Mlima's tusks. This sequence of events has a cumulative impact as we watch Mlima's tusks being unceremoniously smuggled around the world.

The central character of Mlima requires an actor with deft physical invention, sensitivity and power. Mlima is the ultimate elephant, of and for all time. The damage to his body is brutal to witness, the harm done to his eternal soul is what will stay with you. This is a campaigning play of towering emotional heft.

69 ANNA CONSIDERS MARS
Ruben Grijalva (2010)

'If I can convince people to see value in the ugly little things that make more beautiful lives possible... maybe it starts a ripple, that grows into a wave, grows into a tsunami of giving a shit.'

Anna runs the Center for the Preservation of Uncharismatic Species. Right now, she is struggling to find sponsors for Barbara the Pacific mud mantis. Barbara is an insect with an unfortunate problem, her head looks like her anus. '*Do you have something like, er, a panda?*' asks a donor. *Anna Considers Mars* is a fast-moving comedy set in a fast-declining world.

For scientist Anna, in her front-row seat to global mass extinction, Barbara is just as important as a panda. Anna is the champion of species that don't make the headlines, the ugly but important cogs in the ecosystem wheel. Conservation efforts try to focus on entire habitats rather than individual animals, but there's no denying a cute face on a poster can be a greater cash cow than a stick insect. But for every beautiful Spix's macaw there is a less attractive Yunnan Lake newt. For every gorgeous woolly-stalked begonia there is a boring Castle Lake caddisfly. They need just as much help. On top of the difficulties of her day job, Anna is also dealing with her mother's ill health, and her inevitable decline. When a competition is announced to be amongst the first colonists of Mars, Anna can't help but be drawn to the idea of starting again in an undamaged world.

Anna Considers Mars is laugh-out-loud funny, yet it carries an unsettling and pervasive sense that humanity's best days are behind us. It is set in the near future, with half of Anna's life lived in an augmented reality. Everyone wears glasses through which they take phone calls, 'visit' people and conduct their daily lives. These are constantly interrupted by unwanted adverts, inconvenient diary reminders and infomercials. The opportunity for real interaction is as limited as it is sensually deprived. Against this backdrop, the search for the first Mars colonists becomes a symbol of hope, the chance for a new dawn for Anna, and for all of us. Could it be humanity's brave new world?

Bubbling with invention, the play spotlights the thankless battles being fought every day by people like Anna, people who find reasons to stay hopeful each morning. It also subtly raises questions about colonising another planet: is this just a more adventurous way of disowning the problems we have caused on Earth? Watching Anna run out of hope (whilst auditioning for a spot on the rocket) is as moving as it is absurd. Her realisation that humans are probably amongst the least deserving of species worth saving is particularly affecting. If life is found on Mars (thank you, David Bowie) and humans do find a way of colonising the red planet (which is not outside the realms of possibility), who's to say we won't repeat the same expansionist decimation on another delicate environment that we stubbornly neglected to reckon with on Earth?

70 THE TURQUOISE ELEPHANT
Stephen Carleton (2015)

'Kilimanjaro! The snow is melting. I must go!'

If there is an emerging genre we might call eco-farce, this play represents it. *The Turquoise Elephant* is bold, funny and bouncing with fury. An environmental activism group – The Cultural Front for Environmental Preservation – is causing havoc in Melbourne. As the city begins to literally drown in its own shit due to seawater coming into the sewage system, the Cultural Front terrorise the streets, describing the set-piece mayhem they cause as 'art': one particular piece, a government minister's car filled with paint then blown up, is called *Melting Poles*, an homage to Jackson Pollock. A vlogger called Basra is observing the crisis, recording and commenting, safe inside her ivory tower apartment, but doing nothing. Her two aunts are more proactive. Olympia is a disaster tourist, dashing around the globe, desperate to see the last of things before they disappear forever. Her boyfriend builds atmosphere-controlled biodomes in the middle of the desert, calmly making lists of who he might invite inside when the end of the world arrives. Augusta, the other aunt, is a wealthy arts patron and free-market advocate. She buys the Cultural Front's eco-terrorist street work with an eye to a future profit. Into this hyper-vivid cast of characters, Carleton throws identical twins working as domestic help in the family household, hell-bent on ecological vengeance. With this, you have the makings of a gloriously technicolour piece of theatrical chaos.

It is no coincidence that many of the boldest environmental playwriting is coming out of Australia. The nation has been through some of the worst droughts and wildfires in history, yet the dominant political rhetoric remains concern for how to keep the coal fires burning. Stephen Carleton tackles this head-on with this vivid piece of absurdism, as daft as it is angry. Set in the not-too-distant future, in a hotter-than-hell Australia, his critique of those who have power and wealth is as scathing as it is filthy.

The idea of prestige extinction tourism sounds like it should be fiction, but travel companies are indeed offering packages to people willing to pay for 'The Last Polar Bear/Snow in Africa/Galapagos Hawk'-style experiences. The absurdity of these behaviours are dissected beautifully and in riotous fashion in *The Turquoise Elephant*.

71 TRANSMISSIONS IN ADVANCE OF THE SECOND GREAT DYING
Jessica Huang (2020)

'I
have come
to lay
you and
your kind
to rest
For thousands
of revolutions
around her star
you have
absorbed
carbon and
synthesised
oxygen
and you filtered
water
and you
have done so
with such
generosity
and such
care'

The play begins with an elegy for Bennett's seaweed, an algae, now extinct. We watch as a creature called the Being gently thanks this small green organism for its service to the planet on behalf of the multiple species for whom it provided food and shelter, including the humans for whom it played its small but critical part in absorbing our excess carbon.

The Being is a mourner. The Being notices, thanks, and lays to rest all the final species of their kind. The Being has done this throughout history. *'Goodbye West African mud turtle. Goodbye imperial salamander. Goodbye Adriatic sturgeon. Goodbye India monocarpic palm tree. Goodbye striped rocksnail.'* The Being returns again and again throughout the play. It has recently become exhausted by having to acknowledge so many Dyings.

This play tells an epic tale of grief and extinction through the intersecting lives of a collection of human and non-human inhabitants of the Earth. Katrina wants her unborn baby to see snow. The baby's father, Hugo, feels a lack of higher purpose in a world where finding enough food to stay alive is the only useful activity. Carla, recently widowed, finds herself involved in a cosmic relationship with the Being, a creature without age, time, place or gender. These characters travel through the wilderness, weaving in and out of each others' lives, occasionally accompanied by a lynx, the last surviving snowshoe hare, and some locusts.

The Gaia hypothesis suggests that living organisms interact with their surroundings to create a self-regulating, synergistic system that maintains and perpetuates the conditions for mutual life on Earth. *Transmissions in Advance of the Second Great Dying* finds dramatic form for this idea. The Earth speaks in the play, not in words but in sounds: an onstage percussionist is suggested. The Earth's presence is both comforting and unnerving, but it brings a palpable sense of a living entity that has lain dormant for millennia awakening to cradle our return to understanding.

This play is poetic and elegiac, with a mesmeric quality to its circular, rhythmic writing. It also has an inner heartbeat that rages against loss. In grieving the species that have gone it urges us to find an appetite for life in whatever form we can – in food, in sex, in gardening or each other, finding solace in shared hope, powered by shared grief.

LET'S INHERIT THE EARTH 72
Morna Pearson (2010)

'When we said we wanted what's best for the children, we meant our children, not all children.'

Morna Pearson takes the comedy high road and drives straight to the absurdism of our not-too-distant future. This is an anarchic, wryly optimistic, punk-rock musical that lashes together two very different approaches to the climate catastrophe. Do we head into the wilderness and attempt to survive? Or do we blow all of our money on a Caribbean beachfront and sip cocktails while the world goes up in flames?

On a remote mountain in an unnamed country in northern Europe, Scottish couple Jane and Grant bump into Swedish family Josef, Tove and little Lukas. Both families thought they were the only survivors of a flooding and fire catastrophe that has entirely redrawn the shape of the North Sea. Only the highest points of any country are still habitable. Jane and Grant have some Marmite and an inflatable canoe. Josef and Tove have optimism and indefatigable good sense. They keep their son upbeat with a series of white lies.

Meanwhile, somewhere in the Caribbean, the wealthy Highlanders the Mucklefannys and the posh Karrés from Sweden sip cocktails and watch with amazement as an ever-expanding seaweed patch destroys the ocean and eats up the beachfront properties. They remain dismissive of the science even as they slowly metamorphosise, Kafka-style, into sea turtles.

The fate of the super-wealthy in the climate emergency is ripe for a new theatre of the absurd. Morna Pearson places the super-rich on the front line, and waits for the chemical reaction. They reluctantly concede the possibility that their private jets contributed to the beachfront homes now being under seawater, as they casually take a plastic straw from the nostril of a nearby turtle to stir their cocktails. The destruction of their luxury paradise and the disintegration of their human identities is a ghoulishly satisfying watch.

Back on the mountain, the mismatched couples share one excellent joke about Norway, but find very little else in common. Josef and Tove have prepped for this moment, they have their knives, their bottled water, and maps with escape routes. Jane and Grant have stumbled into this situation with their urban sensibilities intact and their laissez-faire attitude ever-ready to wind the Swedes up.

Let's Inherit the Earth is a wild and raucous tragicomedy. It builds to a genuine comic sadness that unsettles. And it has a wonderful final gag about who, in fact, will inherit the Earth.

73 THE BEDBUG
Vladimir Mayakovsky (1929)

'Comrades come closer, don't be frightened – it's quite tame. Come, come, don't be alarmed. Look, it's now going to have what they call a "smoke".'

Prisypkin, a hapless ordinary man, is about to get married. But on the day of his wedding, a fire breaks out and he ends up in the basement of an apartment block as it fills up with water. The severe weather results in him being frozen in a block of ice that remains solid and undiscovered until the building is redeveloped fifty years later. Prisypkin is found by builders who report him to scientists who eagerly defrost him. Miraculously, he survives.

The world Prisypkin wakes up into looks very different, almost a Russian utopia. There is no poverty, no disease, nobody even swears. Could this be happiness? Prisypkin is exhibited in a zoo alongside the bedbug who was defrosted alongside him. However, the scientists' focus turns to the bedbug – *bedbug commonalis* – a species that is now extinct. Realising that this is the last of its kind, the scientists create a specially adapted zoo. They make plans to breed it with *bedbug bourgeoisie*. Prisypkin does not belong in this new world where his fellow proletariat have been eradicated. He sits, forlorn, forgotten, of no interest to anyone.

Written nearly a hundred years ago, Mayakovsky's play was a satire on the social ambitions and philistinism of Soviet Russia. Today, as with many of these plays, the metaphor on which the play is built is too close to fact to resonate with its original intention. Instead, the play starts to take on other meanings, and open itself up to other interpretations.

Scientists around the world are racing to collect DNA from every living plant and creature on the planet, storing genetic material in giant deep-freezes in the hope that in the future extinction may be reversible. This idea – called De-Extinction – has taken on such currency that the International Union for the Conservation of Nature has established a De-Extinction Task Force to provide moral principles for those who may soon be involved in attempts to resuscitate extinct species. How will this code of conduct be drawn and implemented? What political and ethical guidelines will be followed now that reviving life from stuffed specimens is becoming practically possible? Are international efforts being concentrated on saving DNA *rather* than saving species? Is this an admission of defeat?

Perhaps *bedbug commonalis* might prove to be a useful comic template from which to consider some answers to those pressing questions.

EXTINCT 74
April De Angelis (2021)

'I want you to imagine me in a crowd of people. It's 2030. It's midday. The sun is nuclear hot.'

A woman stands alone on stage. She has one hour to change our future. One hour to avert catastrophe.

Addressing the climate emergency head-on, our narrator is an environmental activist who tells us of a time, less than ten years from now, when a pervading sense of panic is causing civil unrest. This is part staged activism, part lecture-demo, part futuristic drama. The play slowly builds a picture of the unignorable facts of the state of the globe in 2021. The writing is at times fact-based and purposeful, at others novelistic and descriptive, its strength coming from the combination of the two. The activist is asking us to imagine, and demanding we recognise the urgency of this moment, and that we finally let the penny drop.

April De Angelis dreams forwards ten years to 2030 and looks at the state Britain will be in: rations, heat that kills, authoritarian government.

All of this is now easier to imagine from Covid lockdowns. She weaves this direct address with a second strand, a narrator who offers us a chain of facts. This is emotionally delivered, never sentimental and always punchy. The third thread of story woven through is the movingly narrated tale of a Bangladeshi family, coping with the floods in the country. Bangladesh has shifted in just one generation towards being more of a floodplain than a country. The death by drowning of members of her family, as told by a returning journalist, is gut-wrenching and real. These three voices complement each other, build in power and create the breadth and detail of the world picture.

The woman on stage ends with a call to do something. She offers us a list of actions that can be taken. This play is urgently put together and was written to be performed before COP26. The deadline ramps up the urgency for us to collectively make a difference. A user-friendly straightforward one-person play with voice-overs. Unapologetically polemical and packed with narrative and suspense, *Extinct* brings the future very close to our current moment, presenting a world not so far removed from the one we are experiencing now.

75 A PLAY FOR THE LIVING IN A TIME OF EXTINCTION Miranda Rose Hall (2020)

'It is really really hard to figure out when, exactly when
a species goes extinct,
because it's not like they just
leave you a note on the kitchen table saying
hey, I'm going extinct,
they just start dying,
and then they're gone – '

A dramaturg, Naomi, walks on stage to apologise that there will be no actual play tonight. Both actors are unexpectedly unavailable. The Zero Ommissions Theater Company were supposed to be putting on a play about climate change. They want their audience to 'WAKE UP!' and are convinced their play, *Climate Beasties*, will do the job. But with the actors indisposed, the evening's entertainment is down to Naomi. Finding it impossible to run the lights and be the stage manager, let alone perform, Naomi abandons the play in favour of a more direct interaction with the audience about the moral repercussions of extinction.

What happened to the little brown bats? Where did the spotted tree frog go? What will happen to *Homo sapiens*? *A Play for the Living* asks us to be honest about our insignificance in relation to the clock of the planet. In doing so it nails the fact that key markers have already passed, that we are looking over our shoulders as much as we are squinting into the sun. It deals with time, with acknowledgment, with responsibility, and the immersion of the human in the web of life, urging us to recognise that we are part of this ecology and not aloof from it.

A formally bold, interactive piece of storytelling, this is a darkly comic and profoundly moving evening of theatre. Written with grace and a light touch, it conjures a powerful emotional response. The play explores how to be human in an era of man-made extinction. Gentle and honest, it suggests that before we can move forward, we must lament what we have lost. This shared action becomes quietly and communally devastating, whilst clarifying the future. There is fuel in grief.

PART 10
FIGHTBACK

Fightback

The question is not if Mother Nature will fight back but how hard. Weeds will grow through cracks in the tarmac, houses will be subsumed by forests, animals will turn motorways into migration routes. Humankind thinks it's in charge but we all know control is an illusion. How will we cope when the forces of the natural world decide enough is enough?

Ecocide is a newly defined term, joining femicide, homicide and genocide on the roster of words we use for murder. It was first coined after the Vietnam War when the United States used chemical weapons to raze both people and environments to the ground. It has since been used to describe harmful activities that cause mass, severe and long-term damage to a particular environment and its inhabitants, whether those be plants, people or animals. At the time of writing, a handful of countries have adopted definitions of ecocide into their national laws. Ecuador has even famously granted Nature its own right to life. Article 71 of Ecuador's national constitution now reads: 'Nature, or Mother Earth, where life occurs and reproduces, has the right of holistic respect of her existence and the maintenance and regeneration of her vital cycles.' The United Nations is currently debating whether to add ecocide to its list of internationally recognised crimes, a decision that could have enormous consequences in the criminal courts.

What follows in this chapter is a collection of plays that honour nature's power. We have included Caryl Churchill's peerless *Far Away,* about an all-out war in nature, and Madeleine George's *Hurricane Diane*, a hilarious imagining of the chaos that ensues when the perfect lawns of suburban America are allowed to rewild. Enjoy the full might of Nature in all her glory.

FAR AWAY 76
Caryl Churchill (2000)

'The elephants have gone over to the Dutch.'

This is arguably the world's most perfectly formed play. It is a cornerstone of twenty-first-century playwriting, influential across generations and geography.

The whole planet is at war. Not just the people, but the animals, the plants, light – everyone and everything is being drawn in to the fighting. Nobody knows who's on whose side, or even why. In an extraordinary three-act play that lasts little over forty-five minutes, Caryl Churchill distils a dystopian crisis into a breathtaking theatrical coup.

In Act One, Young Joan is awake in the dead of night in a remote house in the country. She knows she has seen her uncle loading dead bodies onto a van. Her aunt reassures her it's all perfectly normal. In Act Two, Joan is now grown up and working as a milliner. Her ornate hats are paraded by prisoners, and then burnt, for what purpose she is not sure. In Act Three, Joan's aunt is waiting for her to return. When she finally does, Joan breaks down, overwhelmed that every single thing in nature has now become a threat. Even gravity has been commandeered as a weapon.

And that's it.

But contained within are multitudes.

Caryl Churchill's untrammelled, elegant imagination and her supple language combine in *Far Away* to chilling effect. Young Joan's clarity of perception is befuddled by her aunt. Despite knowing, in the way children do, exactly what is going on, Joan is told comforting stories to reframe evil as normal. Churchill draws out how the wilful suppression of truth – and our willingness to believe a more benign version of events – leads inexorably to totalitarianism.

She links casual disregard for human life, in these night-time acts of ethnic violence, with our daily, incremental chipping away at the environment. In doing so we lose our humanity. Hers is a geopolitical worldview theatricalised with absurdist menace. The play shows how we are sleepwalking into a future where resource wars are the norm and everything in the environment is appropriated by politics, by whichever side gets there first. We are frogs in the warming water, unable to do anything but make hats against an endless cycle of violence we refuse to acknowledge.

We could have chosen any number of Churchill's plays to include in this book: *Light Shining in Buckinghamshire*, *Fen*, *The Skriker*, *Mad Forest*, *A Number*, *Escaped Alone*, *What If If Only*... Caryl Churchill has always known.

77 THE BIRDS Aristophanes (414BC)

'Entities without wings, insubstantial as dreams, you ephemeral things, you human beings:
Turn your minds to our words, our ethereal words, for the words of the birds last forever!'

Imagine a satire in which a falsely sold utopia is labelled cloud-cuckoo-land. Or imagine an allegory in which the triumvirate of humankind, the animal kingdom, and the abstract forces we call 'acts of god', were all in conflict with each other. Imagine a documentary about a world obliterated by deforestation to make warships and overgrazed by herds for soldiers' shoe leather. *The Birds* by Aristophanes is spectacularly all of these things. It is as apt a play as we could find for this moment.

Two Athenians are fed up with life in their great city. Pisthetaerus and Euelpides complain that taxes are too high, society is boring, and Zeus and co., the Olympian gods, demand too much in time and tributes. The men decide to persuade the birds to build a kingdom in the sky and rule the roost instead.

Aristophanes has a bad rep. His plays are seen as diffuse, rambling and just not as funny as they must once have been. Whether this is a fault of translation or taste, maybe the time has come for a fresh look at his writing. The environments he was building inside his plays and the hubris he was mocking are now horribly familiar. Athens' decline is attributed in part to its leaders' neglect of the wider ecology of Greece.

The centre sucked all goodness out of the periphery to the detriment of both. Aristophanes knew what he was lampooning.

The Birds is a casting director's dream. The king is a hoopoe bird called Tereus who, having discussed the Athenians' idea with his nightingale wife, agrees to take on their request and establish a new natural hierarchy. The nightingale sings to call all of the birds together, those of the seas and the land, the fields and the mountains. A bird parliament is held to plan this new kingdom. But not all of the birds are so trusting. Humankind has historically been responsible for trapping and caging, eating and stuffing. Why should it be trusted now?

Pisthetaerus persuades them otherwise. He reminds them that the birds were the original gods until they had their privileges stolen. It is also pointed out how helpful birds are to humans, in keeping insects in check and acting as early warning signals for changes in weather and seasons. Why can't they find a happy symbiosis? The birds agree to the plan, but only if the Athenians dress themselves as birds too and join them wholeheartedly in their takeover of the world. This, the humans are happy to do. After some discussion they agree to name the new realm of the birds Νεφελοκοκκυγία, which translates as cloud-cuckoo-land.

The play's absurdity gives it a startling prescience. It is beautiful in its depiction of the grace of the birds, of their species-specific behaviours, of their eternal wisdom. It is sharp on the urbane solipsism of the Athenians, and the pure, raging, jealous energy of the Olympian gods. *The Birds* makes it abundantly clear just how mankind stacks the odds against nature. Maybe we should check over our shoulders, to see who the gods are coming for, now we have made room for vengeance. We need a play that humbles us, that checks our inordinate appetite for control, and that lets us laugh at ourselves as we recognise our flaws. The birds, double-crossed by a power-hungry human yet again, are actually not too bothered. They know, deep down, that human life has nothing on their timeless, soaring immensity.

78 SEA FRET
Tallulah Brown (2017)

'Loving where you live with every bone in your body has got to count for something.'

Set on a Suffolk beach as the sea creeps ever nearer, this is a dynamic coming-of-age love story. The love in question is twofold: the deep friendship between best pals Ruby and Lucy, which is about to be ruptured with Lucy's imminent move to university; and the love of place, their profound connection with the coastline they have always called home.

A sea fret is a wet mist or haze coming in off the sea. This specific coastal phenomenon is just one of the many things Ruby and Lucy love about their patch, along with its phosphorescence, its particularly salty salt and the wondrously shaped flints they find underfoot. Here is where they made all their teenage memories, drunken fumbles, wild parties, deep confessions. Their teenage dialogue bites with authenticity and specificity, the writer Tallulah Brown creating a parlance unique to these two individuals, every line imbued with their shared jokes and memories.

Ruby lives with her drunken, stoned dad, a benign but useless father figure. Their home is a ramshackle pillbox under constant threat from the encroaching sea. Lucy lives with her practical, purposeful mum, further inland, safe from the waves. The play's four scenes are structured using the Environment Agency's terminology for different approaches to coastal erosion, a language that is a peculiar mix of military and bureaucratic terminology: 'Advance the Line', 'No Active Intervention', 'Hold the Line' and 'Managed Realignment'. Ruby and Lucy would prefer to have one last mad summer on the beach before Lucy heads off to university. There is sex to be had and drugs to be taken. But the community must make urgent decisions about its future before heading into another winter, and Ruby and Lucy can be young carefree women no longer.

The UK has around 2,000 miles of actively eroding coastline, with a further 1,500 miles shored up by artificial barriers, the longest in Europe.

With limited options available to them, the community in Tallulah Brown's play must make a difficult choice. A town-hall meeting is planned and Ruby and Lucy, along with their parents, are expected to attend. Speeches must be made and homes protected. Reality hits as the two households must make difficult decisions about whether to let nature take its course or whether to intervene. Communities up and down the east coast, and around the world, face this exact, impossible dilemma.

HURRICANE DIANE 79
Madeleine George (2019)

'You've been busy, haven't you? Mining and stripping and slashing and burning and generally despoiling the green earth that gave you life.'

The Greek god Dionysus – reincarnated as a gender-queer landscape gardener called Diane – comes down to Earth in a last-ditch attempt to help us save ourselves. Starting with four posh women in suburban Upstate New York, her plan is to start a rewilding revolution. Tear up those perfect lawns! Get rid of that imported hedging! Ditch the subtle-colour-palette planting! Let your weeds self-seed, make space for native species, and embrace the overgrown. And have a whole load of orgasmic organic sex on the way. This hilarious, rebellious, Bacchanalian play-orgy shows what fun we can have if we'd just lighten the fuck up and let nature do its thing.

How many hours are spent each week in gardens around the world by people mowing their lawns? Keep it down, make it neat, fit in with the neighbours. The social conformity and natural repression represented in lawn mowing is exploited to ruthless comic effect in George's raucous play. The lawn represents discipline, control, civilisation – as opposed to naturalisation, release and freedom. *Hurricane Diane* is a Greek comedy mixed with *Desperate Housewives*, with a dash of *The L Word.* In a mixture of casual realism and cosmic charge, it eviscerates the blind eye we all turn to the climate crisis in our quest to keep up with the Joneses and not let the neighbourhood down.

Have the lawn wars started in your neighbourhood yet? Just see what happens when you let your grass grow to knee height. 'It lowers the house prices.' 'Weeds blow into my garden.' 'It looks out of place.' Diane represents all that is good about nature, and all that could be good about humans, if we could only embrace the Earth, and each other, with openness and ecstasy. There, might we find salvation. As Diane/ Dionysus puts it in the play: '*You don't know what time it is on the cosmic clock. How could you? With your bird lives, your fruit-fly lives, hatching and feeding and breeding and dying, all in the blink of a god's eye? So, let me tell you what time it is. It's eleven fucking forty-five.*'

Uproariously funny and slyly sobering, *Hurricane Diane* is a welcome storm.

80 NORTH COUNTRY
Tajinder Singh Hayer (2016)

'I tell them that's what's going to keep us alive. Connection. When things got bad, it wasn't just because people got ill, it's because they stopped looking out for each other.'

North County was reviewed as 'science fiction' when it was first performed. It follows three characters in the aftermath of a plague that wiped out much of the population. In the short time since it was written, the play now feels a lot closer to lived experience.

'Plague' still feels like a peculiarly biblical word, summoning images of locusts, carrying connotations of vengeance on the ungodly by furious deities. Historical perspective, however, can change our lens. The Black Death, the Spanish flu, Covid-19, they are all given potency by travellers. The mass of soldiers returning home from the First World War was the greatest single migration of people in our global history and stoked the swift spread of the Spanish flu. It's this ease of movement and pace of transmission that is assumed to have led to the plague in Tajinder Singh Hayer's absorbing play.

North Country is set in Bradford in the north of England. Harvinder, Nusrat and Alleyne have all survived the plague that has swept the country. Thrown together by their unexplained immunity they must now cooperate. The landscape is bleak and technology-less. Putting aside their differences is the only way to survive. The play cannily jumps across three different timelines. We see them in the immediate aftermath of the plague, thirty years later, and then thirty years later still. Throughout the characters re-evaluate their abilities and refocus their expectations. Bartering becomes the key currency, of stuff and of knowledge. Alliances are transferred, positions shift, and wiliness proves to be the most reliable survival tactic.

Modern plagues, such as SARS and swine flu, tend to flicker to life, spike and then die. Potent cocktails of drugs and social exigencies smother their possibilities. Until the next time. At the time of writing, Covid-19 is proving to be a plague that has broken ranks. It is sweeping the world, causing death wherever it goes. Scientists argue plagues are becoming increasingly common because of habitat destruction. As humans come into closer contact with animals and plants from previously undisturbed environments, viruses jump from creature to creature, with no natural evolutionary controls. As the planet heats up and permafrost thaws, more dormant bacteria are set to be released with similarly unknown consequences.

North Country is a fascinating – and useful – guide to how to thrive when the worst happens. The three characters are from different classes and cultural backgrounds. Harvinder is middle-class, overeducated and well-brought-up. Nusrat is younger, a working-class British Pakistani. Alleyne is a white farmer. Harvinder has medicine, Alleyne has animals and a gun. Nusrat has guile and compassion. As the play follows their shifting understandings and allegiances, it reveals a wider story about identity and the process of rebuilding – or building anew – a national identity after a time of crisis.

81 WOLVES ARE COMING FOR YOU
Joel Horwood (2017)

'No one knows but this has always been a place for stories.'

Someone in the village has seen a wolf. How did it get here? Where did it come from? More importantly, what can be done about it? *Wolves Are Coming for You* asks us to consider just how much wild we think we can live with.

Joel Horwood's play requires a dynamic mode of storytelling. The play is set in a remote community populated by a variety of characters, each one expressing a different viewpoint of the wolf, of their neighbours, and, revealingly, of themselves. A vivid community wrestling with its internal demons as well as its external ones is beautifully conjured. Rumours intensify and the tension in the village ratchets up. The encroaching wilderness outside seems to spur the villagers to unravel inside. Home truths are told, long-held grievances aired. The fear of the unknown plays havoc with these imaginations.

The term 'wilding' or 'rewilding' is understood in multiple ways by different environmental stakeholders. For some, it means simple non-intervention, allowing a piece of land to remain untended. For others, it is reintroduction, facilitating the return of a particular species. Within this large span of opinion and understanding there is common ground. Advocates claim the return of long-absent flora and fauna provides much-missed benefits to struggling ecosystems and encourages biodiversity. Opponents worry about damage to farming and the risks of unpredictable human/animal interactions.

In theatrical terms, the 'wild' has long been a metaphor for human fears about outside threats to an established way of life. This fear has been focused historically on excluding particular groups of people on thinly disguised premises, or on unwanted changes to cherished landscapes. The language of 'native' and 'alien', 'natural' and 'introduced', 'roaming' or 'managed' all point to similar fears running through our environmental policies, whether that is control over people, plants or animals. The evidence points to the universal benefits of allowing nature and animals to restore their own agency, with humans as an enabling rather than corralling force. Equally, with national borders becoming arbitrary in the face of global movement of peoples and animals, how much control are we willing to cede? This may become the key question of the next decade. Joel Horwood's suspense-fuelled drama forces us to confront the tiny fearful ache inside.

HUMAN ANIMALS 82
Stef Smith (2016)

'Don't go burying wild animals in my garden… or at least ask for permission first.'

Nature is out of control. Foxes are running riot in the streets. Rats are making themselves at home, in people's homes. And the bloody pigeons won't stop multiplying. Something's got to be done. Stef Smith's inventive play imagines an urban eco-crisis in the form of an animal infestation.

The play is a fantastically clever twist on extinction, taking abundance rather than paucity as its starting point. It depicts a London overrun by rapidly increasing numbers of a few urban species. As their populations spiral, humans are pushed to their limits, revealing just how thin the veneer of civilisation really is. Logic and reason are replaced by guile, cunning and irrationality as fear grips society. Who is going to do something about it? At what point should animal rights be shelved in favour of population control? And who exactly should do the culling? Six interlinked characters – young lovers Lisa and Jamie, mother and daughter Nancy and Alex, alongside their neighbour John and his new partner Si – all find themselves on different sides of the debate.

Megacities typically have over ten million inhabitants. There are currently around thirty in the world, mostly found in South America and Asia, with the number set to expand in the coming decade. Problems of pollution, housing, crime, congestion and waste abound. It seems incredible that two hundred years ago less than 3% of the world's population lived in cities. Today the number is 60%. The default response of most cities managing their overcrowding issues is to expand. Whether up, down, out or across, creating more space inevitably means encroaching. By asking us to experience animals expanding numerically and relentlessly into 'our' territories, Stef Smith flips this on its head. She acknowledges the agency and autonomy of non-human animals, forcing us to experience it from the other side.

The play shows how quickly a situation can develop from minor inconvenience into major crisis. With rumours of infection and disease spreading, advice leaflets are soon out of date. Road closures work for a while but protests against curfews further escalate the tension. Society is on the brink of breakdown. Soon all other options are off the table, as unseen authorities decide universal mass culling is the only option. A bloody end ensues. The fear of the other and the fear of abundance is fabulously dramatised in this terrifying parable.

83 A FABLE FOR NOW

Wei Yu-Chia (2018)

Translated by Jeremy Tiang

'Hello? World Leaders' Summit Emergency Hotline. Is this urgent? What? Too noisy here. Hang on.'

A bereaved polar bear is on a fag break from her work in a fish canning factory.

A duck and a panda discuss how to best perform being professional zoo animals.

A group of infants play at being world leaders (or are they actually world leaders?).

A genetically modified chicken packages itself into a tin.

Two soldiers realise the birthmarks on their bodies come together to make a love heart.

An immortal tortoise proves itself otherwise.

The last man alive lectures to no one on a desert island, the detritus of the play at his feet.

In seven unforgettable scenes, Wei Yu-Chia's unique imagination presents different versions of the end of the world.

Inspired by a random assortment of news reports and images – an emaciated polar bear floating on a piece of ice, a giant yellow duck art installation travelling the world, North and South Korea resuming old hostilities, hysteria in Taiwan at the birth of a panda at the zoo, an outbreak of poisoning in a food-processing factory – *A Fable for Now* uses a collage format to capture the kaleidoscopic information overload of the crazy times we are living through.

The pictures we see all day long signifying environmental catastrophe can be hard to unsee. The dead albatross with Lego in its stomach, the barefooted boys playing football on the world's largest rubbish dump, the forest fires inside the Arctic Circle. Grotesque and absurd, these macabre images are in danger of becoming normalised, their ubiquity diluting their power to shock. The play dares us to look again, and this time look beyond.

The speaking animals and the wild images are the theatrical fun of this play, but its quieter power lies in its potent language and the writer's innate understanding that incongruity gives us access to the richest form of deep thinking. Her masterstroke is to let the sublime and the ridiculous come together in the universal language of fable.

A surreal apocalypse on one hand, a simple folk tale on the other, *A Fable for Now* is a mischievous and gorgeously welcome shock to our nearly numb sensibilities.

PART 11
AFTERMATH

Aftermath

The concept of deep time helps us think on a scale beyond everyday human imagining. It encapsulates geological events stretching back millions of years, back to the formation of the Earth itself. This is a timescale unimaginable to most humans, who consider time relative to the span of themselves, their ancestors and their offspring. The idea of deep time can be useful for theatre-makers. It's a concept that can be projected forwards, beyond the immediate crisis, giving us the perspective of future civilisations looking back at our current actions. In theatre, 'time + place + action' are our raw materials, so how can we distort time in the fabric of plays to give it more weight, presence and focus? Plays in this chapter have taken time as their central conceit and wrangled new thoughts from this most invisible component of our storytelling.

It is technically difficult for a dramatist to ask us to anticipate regret for our current passivity, but there is a need for us to experience this present moment vividly enough to re-examine our actions. It requires a great deal of invention from a writer to reconfigure the present moment and make it feel newly potent. Each of these plays manages the Brechtian trick of making the present moment feel *just unfamiliar enough* for us to see it anew and understand it afresh.

We have tried to steer away from straightforward dystopian drama (as enjoyable as that can sometimes be) and instead we selected plays that engage with an imagined post-crisis world with wit and energy. These plays – ranging from Thornton Wilder's totemic *The Skin of Our Teeth* to Jackie Sibblies Drury's mischievous *Social Creatures* – step forward maybe a generation, an epoch or an eternity, or even slip into a liminal space where time is uncertain. They each in their different ways tempt us away from our cosy present-day sensibility and ask us to reconceive our place in time.

84 THE SKIN OF OUR TEETH
Thornton Wilder (1942)

'My nerves can't stand it. But if you have any ideas about improving this crazy old world, I'm with you. I really am.'

It isn't easy to break every single theatrical convention in one play, but Thornton Wilder certainly had a go. *The Skin of Our Teeth* is genre-defying in form (one character frequently steps out of the narrative to express her doubts about the quality of the play) and kaleidoscopic in content (a dinosaur, the Old Testament prophet Moses, and the poet Homer all make appearances). The play is a soaring (and roaring) account of humanity's time on Earth.

The Antrobus family, who take their name from the Greek word '*anthropos*', meaning 'man', are the aptly named central characters, steering the audience from the Ice Age into the Anthropocene (our current geological era). The family are thinly disguised biblical characters, yoked together to represent an average household who must navigate everything an unsettled world can throw at them. The play addresses, amongst other things, the miracle of evolution, the possibility of extinction, and the ever-present chaos beneath the veneer of civilisation. The Antrobus family live extended lives across eons, including burning everything they own to survive the Ice Age, turning up at a political convention for the Fraternal Order of Mammals on the eve of the flood, and a final act which sees them hunkered down while a long and devastating war finally comes to an end.

The play's great strength is its original and unruly chronology. It is simultaneously set now, in the past, and in the future, freewheeling through a flexible, layered concept of time. *The Skin of Our Teeth* begins with the invention of the wheel and ends in the aftermath of global conflict, via a flooding event that Noah would recognise. The sense of foreboding in the run-up to the flood is particularly potent; the weather changes from sunshine to hurricane to deluge, while a fortune-teller who attempts to warn the characters that something bad is coming is routinely ignored.

You can sense how much fun Wilder had writing the play; his wit bounds across every page, and the dialogue runs with a free-form glee that at times threatens to overwhelm the action. To keep his audience on board, Sabine, the housekeeper, directly asks the audience if they have any idea what the hell is going on. *The Skin of Our Teeth* is a gift for actors and creative teams, allowing a no-holds-barred approach to production. Did we mention there was a woolly mammoth?

Wilder wrote this play during the Second World War. The flair sits on top of a burning anger that we almost destroyed everything. The play reveals Wilder's deep affection for how ordinary people somehow survive the extraordinary chaos of global events. His is a hopeful worldview, a belief that somehow we will find a way, and that we are better than the chaos. And when we do find a way to survive the present moment, the title sums up how.

85 MR. BURNS, A POST-ELECTRIC PLAY Anne Washburn (2012)

'Yes, yes, yes that's right, right? And Lisa is saying, like, that's really funny Bart why aren't you laughing or something and Bart is full of gloom because he's depressed because someone is trying to kill him...'

Why do we need stories? *Mr. Burns, a Post-Electric Play* poses this question in three startling and audacious acts. An unnamed disaster has made generating electricity impossible, plunging the characters back into the Dark Ages. From there they build forward with half-remembered snippets of a television programme, shards of memory collectively reassembled, and out of this simple shared experience a new culture is born. Storytelling becomes the binding that knits together this blasted community.

The play (with original music by Michael Friedman) tells the story of a group of survivors who spend their time recalling and retelling a particular episode of the American animated TV show, *The Simpsons*. The episode they choose is 'Cape Fear', which is itself an animated retelling of the movie *Cape Fear*. Seven years later, the characters are now a theatre company specialising in performing *Simpsons* episodes; we see them sophisticate their theatrical retelling. Their aesthetic is inventive and witty in ways the troupe are competitively proud of. In this second act, several years after the initial disaster, there is a booming business in cultural reconstructionism, but the real competition is between those who do *The Simpsons*. The final section of the play is set a further seventy-five years in the future with the same episode being performed, now as well known and as culturally important as a Greek myth. This third act takes on the quality of a musical pageant, adapted to fit the cultural needs of this new society who are still coming to terms with the near-extinction of humanity nearly a century before.

Washburn's postmodern mash-up of popular culture, theatrical know-how and musical invention is hugely entertaining. It also excoriates capitalism and consumerism (of which *The Simpsons* character Mr Burns is emblematic) creating a powerful modern morality tale. This idea is developed in the second act as the characters barter for the right to use certain lines from *The Simpsons* canon. It couldn't be more appropriate or ironic that copyright laws have survived the apocalypse when artists and authors have not. The play dramatises how culture becomes commoditised, but also acknowledges and rightly celebrates the way stories and shared cultural artefacts bind communities together.

All this action plays out in a context that is genuinely unsettling. The characters do not know what happened or who else might still be alive out there. They are afraid of toxic exposure, violence and what else the future may hold. They have all been through the trauma of the end of civilisation without a template for moving forward. When an unknown survivor suddenly stumbles upon the group gathered around the campfire, everyone asks about lost loved ones in the hope someone may have heard something, that loved ones may have survived. It is a bleak and sobering moment.

Anne Washburn's play explodes with originality and ambition, rejecting worthiness and replacing it with a blast of dauntless good humour and a ferocious intellect. Perhaps in the future a group of survivors will come together and find a shared humanity in recalling and retelling *Mr. Burns, a Post-Electric Play.*

86 LUDIC PROXY
Aya Ogawa (2021)

With Japanese translation by the writer
and Russian translation by Ilya Khodosh

'Since last week I've been having this dream. I'm walking on a soft sandy beach by myself, next to a beautiful blue sea. But soon I notice that in this blue sea, there's not a single fish or living creature. And what I thought was beautiful white sand isn't sand at all, but ashes.'

先週から同じ夢をみてるのよね。誰もいない、きれいな青い海があって、その柔らかい砂の上を私一人で歩いているの。日差しが暖かくて、、、波の音を聞きながら、お腹の 中の赤ちゃんのことを考えてるの。でもふと気がつくの。 青い海の中には、魚とか、生き物が全然いないことに。それで分かるの。私が奇麗な砂だって思っていたのは、実 は灰だったって。

This astonishing debut is a multilingual, multimedia piece of immersive theatre. The play takes place in three damaged landscapes. Act One is set in the near past, following the story of a woman who survived the Chernobyl nuclear explosion. Later in life she becomes obsessed by a video game that takes her hometown of Pripyat (abandoned after the disaster) as its setting. She plays the game, compelled to retrace her steps and attempt to rewrite her history. The dialogue is in Russian, with English subtitles.

Act Two is in the present and offers the audience control of the narrative. Through a daring multiple-choice structure, the writer asks the audience to imagine the choices facing a pregnant woman living on the outskirts of the Fukishima evacuation zone, torn between abandoning her home town and saving herself and her child. The dialogue is in Japanese, with English subtitles.

The final act is set in the near future where all humans live underground, unable to venture above due to toxic levels of pollution in the atmosphere. The world below ground is possible only thanks to pervasive and invasive technology. In a hospital, a woman holds her mother's hand as she dies. But the contact is virtual, the last moments recorded. Refusing the 'Afterlife session' on offer, the woman instead goes up to the surface to see the remnants of real organic life. The dialogue is in English.

This unsettling play asks the audience to think about what it might mean to live in a dangerously polluted atmosphere. Are we better off living in sealed bubbles or even inside virtual worlds? Where might humans best survive?

The story moves between real and virtual words with dazzling ease. The use of multiple locations, languages and theatrical forms has a powerfully disorientating effect, demanding the audience keep up. Throughout the play, a present-day Nina from Chekhov's *The Seagull* takes up a haunting refrain, the nineteenth-century lines now taking on an eerily prescient power: '*Thousands of centuries have passed since this earth bore any living being on its bosom...*'

The spectre of nuclear fallout provides a chilling backdrop to the action. Chernobyl will not be safe for another 20,000 years. The site is now part wasteland, part uncontrolled biological experiment, part disaster-tourism hotspot. The question of where and how to store nuclear waste from power stations is an equally long-term problem. Scientists talk of 'geological disposal', meaning it takes geological eons of time for radioactive isotopes to decay. Huge underground caverns are being created to house high-level nuclear waste. The builders who work on these sites call them 'post-human architecture', structures designed to keep their contents safe for tens of thousands of years, well outliving the human race. There is much debate about what might be the appropriate signage to warn any future beings from inadvertently opening the door. As the nature writer Robert Macfarlane questions, what universal symbol would you choose for 'DO NOT ENTER HERE'?

Ludic Proxy crosses continents, timescales and theatrical genres to pose similarly uncomfortable questions. If we think nuclear power is a 'clean' form of energy, are we prepared to live with the consequences? The end of the play sees time and stories collapsing in on each other, the characters walking towards an uncertain future through multiple layers of reality.

87 HAPPY DAYS
Samuel Beckett (1948)

'Perhaps some day the earth will yield and let me go.'

In *Happy Days*, Samuel Beckett reduces the story of a life to a mound of detritus that his central character, Winnie, is being buried beneath. Her possessions define her. The line between where she ends and where her things begin is blurred. Winnie is embedded in her own belongings where she cheerfully sits out the end of her days. Her routines involve picking through her handbag to check her bits are still there, her lipstick and toothbrush taking on symbolic meaning. Beckett observes how the ritual of ownership defines the way we build our lives and our sense of self. Items are imbued with meaning, the look and feel of a particular piece acting as a trigger for a memory or story. But then the cherished object joins the mound of detritus, and loses definition, swallowed by the earth.

The average person owns about 300,000 items across their lifetime. These include large purchases like sofas to smaller ones like books. The majority of our possessions will end up in landfill; less than half of waste in the UK is currently recycled. (So either keep this book on your shelf *forever* or pass it on to someone else who will do so when you are done with it.) Rubbish is one of the most unpleasant – and untold – problems of the modern world. Almost every country on Earth has run out of space to dispose of their waste, with burial or burning the only viable options. Waste finds its way into the sea, creating flotillas of household rubbish; or even into space, experimentally rocketed out of the atmosphere to become the galaxy's problem rather than ours.

For Beckett, language is the balm, the only thing of value we have left. Many of his plays could have made it in to this selection – *Krapp's Last Tape, Endgame, Waiting for Godot.* He knew that it would be words and words alone at the very end. He also knew that time is a dramatist's gift and he could manipulate it like no one else.

In *Happy Days,* Winnie swings between eternal optimism and darkest despair. Brightly telling funny stories packed with innuendo, while the gun in her handbag hints at the option of suicide. This combination of farce and tragedy captures the tension of human resilience. Winnie beats back the terror of death with laughs and memories and words, mostly to herself, occasionally to her unseen, monosyllabic husband. Only fleetingly does she allow the fear to break through, knowing that the anticipation of the worst is worse than the worst itself.

The play calls for a deluge of light, so bright that Winnie has to use her parasol to protect herself. During the play the Earth's fierce final sunlit days beat down, before finally darkness engulfs them. It is both a cheerful and devastating image of the end. Beckett is a true seer of the aftermath.

88 WHAT USE ARE FLOWERS? Lorraine Hansberry (1972)

'You eat raw meat, don't know fire and are unfamiliar with the simplest implement of civilisation. And you are prelingual. '

In *What Use Are Flowers?*, playwright Lorraine Hansberry imagines the planet after the end of the civilised world. In this one-act play, an elderly man lives as a hermit in a forest for decades, rejecting society and its ills. He emerges to discover that he is the only adult left alive in the world.

Hansberry does not specify what disaster has befallen the planet. When she wrote the play, the greatest fear was nuclear holocaust. The play, however, leaves the context enticingly open. It could easily be a political collapse, an environmental apocalypse or an all-consuming war. All we know is that civilisation has entirely vanished, and in its aftermath a few nameless feral children happen upon the hermit.

The old man tasks himself with describing to them what life used to be like, and in doing so he teaches them about what he has lost, and about a way of life they have never known. The man becomes aware of the things that are worth protecting and sustaining for the world yet to come. His task is to connect with these children so they can learn to value the things he values. The hermit decides that what sticks in his memory must be what matters most, surely? His first choice is communication, and so he begins to teach the children language. He gives them names. He teaches them the poems of Walt Whitman. He teaches one boy to play Beethoven. Working with the earth comes next, using natural resources such as clay to make pots for containing and transporting water and food. He teaches them the verb 'to use' and develops their need for practical thinking. This quickly becomes a utilitarian mindset which itself soon becomes philistine. To counter this narrowness, the idea of beauty arises, which, like Socrates, the old man finds difficult to explain. The confused children cannot imagine what this sense of wonder he describes could possibly have been used for. What 'use' are flowers? What is practical and therefore essential about them?

Hansberry successfully creates a counter-intuitive longing for the things we already have, and a need to value and protect them. It is a gentle, deftly written plea.

89 I AM THE WIND
Jon Fosse (2011)

'I'm a concrete wall that's breaking into pieces.'

The action of this play takes place on a boat. Two characters named 'The One' and 'The Other' are stranded on a shore, for reasons never made explicit. The One tends towards pessimism and The Other towards altruistic optimism. This relationship is held in a gentle, loving tension throughout. We experience the shock and paralysis of an extended silent opening. Something grave has happened. Are these men the only two left alive?

The One suggests they take their boat and head for the open ocean. Suddenly the wind gathers, the sea swells and this odyssey into the unknown becomes a battle of outlooks and expectations, between the human instinct for survival and the hypnotic allure of a dark and forbidding sea. The One eventually joins with nature and becomes the wind. The Other is left achingly alone, adrift and in grief.

Playwright Jon Fosse makes no concession to conventional plot or subject matter. There is a Beckettian bleakness and an adroitly established tension between these essentialist worldviews. (A precis could read: 'To Be and Not To Be went out in a boat together. Only To Be came back.') We learn little background detail about the characters beyond the fact that The One is a confident sailor suffering from depression and The Other is a novice onboard. In some moments they say very little, giving the audience space to wonder what has led them to this moment. At others, they communicate effortfully, sharing their frustrations at the inadequacy of language to express the immensity of their inner feelings.

The play reaches its devastating conclusion when The One falls (or chooses to fall) overboard. Deliberately resisting all of The Other's efforts at rescue, The One lets go. Taking place in an environment where there is little chance of finding safety, and in which dependence and loneliness are so vividly sculpted, it is hard not to find resonances about our climate emergency. It is easy to imagine the play occurring in the context of melting ice caps and rising ocean levels. The sea is described as freezing. Seen in this light, The One's depression takes on an unexpected quality.

Ecological grief is a modern condition describing the psychological pain people experience due to witnessing environmental loss. First articulated by biologists working to chart the decline in coral life on the Great Barrier Reef, it has become common and acute in scientists who are asked to observe the loss of species or ecosystems. More recently, this condition has been observed in Indigenous Peoples witnessing the destruction of ancestral habitats as well as being linked to ecological anxiety, felt by many people in different countries around the world who struggle to find reasons to keep living in a time of such uncertainty.

And yet this enigmatic play carries hope. At times the characters veer into comedy, the absurdity of their situation as funny as it is terrifying. At others they demonstrate incredible affection for each other, a shared understanding that goes beyond words. The taut, suspenseful writing keeps their precarious position anchored in psychological truth, while all around them allegorical images and fundamental questions rise to the surface, only to be chased through the vastness by the next wave.

However you choose to interpret this elusive play, it has a sparse, poetic quality that carries both real and metaphorical power. As a study of two people trapped in an extreme situation, battling insurmountable odds, it speaks to human courage and fear, love and loneliness. As a parable of being adrift in a dying world, it celebrates the fundamental instinct for survival as well as acknowledging the deep understanding that we are part of nature, to be 'the wind' from where we came. The ending is both moving and cathartic.

90 THE TRIALS
Dawn King (2021)

'You have stolen my childhood with your empty words.'

There is hope that the problems of present humans might be alleviated by future generations. Many people born into this era of unprecedented climate unpredictability have so far proved themselves far more adept at dealing with its issues and articulating its complex challenges than their parents. Generation Greta might yet save us all.

But is it fair to load that responsibility on to their shoulders? Are we abdicating responsibility? Dawn King's play asks exactly this by imagining a future post-crisis government inviting young people to put their elders on trial. The play presents a form of justice whilst asking is this fair? Did they have it coming? Or is this retaliation couched as justice?

There is a generational divide in levels of concern for the climate emergency. Surveys show those under thirty tend to rate it as their highest worry, while those over sixty rate it as their lowest. Global youth have proved adept at shifting the climate crisis up the political agenda. Awareness and urgency are increasing. There is a sense that this generation can see, in ways their elders cannot, just what needs to be done – how much, how far, how fast.

Yet with fearlessness comes anger. *The Trials* is set in a world that is depleted; no snow, no ice and no coral reefs. A group of young jurors in their late teens sit making decisions about whether to mete out the death penalty to older individuals who did little or nothing. It is hard not to recognise ourselves in the honesty of the defendants who were doing what they thought was enough.

The Trials is structured around three particular defendants. These each make statements explaining their life choices, admitting their flaws and protesting their innocence. They are followed by three sets of deliberations as the jurors debate each defendant's level of culpability. The power of the play comes from making the audience feel that their present-day lifestyles will, in fact, be deemed criminally negligent in the future. This sense of complicity might be uncomfortable viewing for some, but perhaps a bit of squirming might not be a bad thing for others. Which side of history might we prefer to be on? What do we want our legacy to be? The twist of the play is – of course – that one of the jurors is related to the final defendant. Can fair judgement be passed?

The trials are constructed like those in Nuremberg. King is making the comparison that they too involved prosecution for crimes that were not on the statute books at the time they were committed. The play therefore poses interesting questions about retrospective justice. How should governments legislate for climate disturbance? What exactly does guilt mean and who can be deemed culpable? Did the fault lie with those in charge of big corporations? Or are those who drove a car, enjoyed foreign holidays and didn't quite do enough equally to blame?

In 2021, the legalese of this issue is swiftly moving up the international agenda. Bolivia has created a precedent for 'crimes against Mother Earth'. Lawyers serving the interests of people living on low-lying islands – on the front line of already-rising sea levels – are expected to bring test cases against multinationals within a number of years. The sticking point appears to be whether to bring charges based on crimes against the climate or crimes against humanity. (Or are they one and the same thing?) Some large companies are reportedly beginning to set aside considerable funds in anticipation of future compensation claims. But can an offer of reparation come without some acceptance of responsibility?

On an individual level, the concept of intergenerational equity is a widely recognised principle of environmental law. This states that accounting for the preservation of the environment and its resources for the benefit of future generations must be at the heart of international policy. But as of yet this 'principle' does not have legal parameters.

The Trials is a provocative and potentially divisive piece of writing. That is the point. Imagine being a young person in an audience of older people watching the play? Or an older person surrounded by those younger than yourself? With courtroom staging options available, any production has ample opportunity to create a powerfully charged atmosphere, with scope for follow-up debate and conversation.

91 SOCIAL CREATURES
Jackie Sibblies Drury (2013)

'The fiction of individual responsibility died a long time ago.'

This is a very funny, acutely observed social comedy/zombie thriller built on a cracker of a thesis. It just happens to be set in a post-apocalyptic America.

A couple are having a row in the basement of their apartment block because the generator doesn't work and they don't know how to fix it and he doesn't trust that she is reading the manual right and she can't stand that he thinks she can't even read a manual right and they are directing all of their pissy angst at the generator rather than at each other because they are so, so practised at passive-aggressively going for each other's throats. And we learn that they are trying to switch the generator on because things have got really bad, and people only need a generator as back-up when things have got really bad, and we start to realise just how bad things have got when he hits the generator so hard with the wrench that it starts to work and the lights come on and we see that we are, in fact, in a dilapidated theatre, all of us together, and some terrible crisis has befallen mankind.

This theatre is a place of refuge (point taken) and there are only seven people left alive. One of them, Mr Smith, has gone outside, leaving six. This is bad. No one goes outside. No one really knows what is outside the walls of this theatre (point taken). There are shared, stacked tins of food, everything counted and labelled, and allocated for future use by specific persons. There is private time when each person gets a few minutes apart from the others. No one uses their own names. We have the Smiths, Joneses and Wilsons. And the singular Mr Johnson. We learn about their histories from personal videos they record between scenes, individually, as a record of what has been lost, and what they survived to get here. The play takes place in real time, with the videos compressing the past into this hyper-vivid present.

A stranger comes in. A black person, Mr Brown. The only black person in the theatre (point taken). They lock him in a plastic cubicle in case he is infected (so many points taken). He casually observes the situation and makes one of the best jokes in theatre history. This planet-wide crisis of zombie-fied cannibalism and destruction is as a direct result of unbridled white privilege having finally exploded all over itself. Then Mrs Smith, whose partner disappeared at the top of the evening, shows symptoms of the disease previously discussed and is locked in the cubicle with Mr Brown. She eats him. It gets gory. *As he predicted.* Then they leave us in this dilapidated theatre with our understanding of ourselves sharpened.

This play was written nearly ten years ago and has become ever-more pertinent. The anxieties of the outside world, the sense of impending disaster, the constant vigilance for 'symptoms', the being locked in, are all very twitchily post-Covid funny. This writer has vision.

PART 12
HOPE

Hope

Many of you will have to come to this chapter first. Flicking through a book like this, it's where most of us would want to begin. Hope sets up quite a promise.

In Greek mythology, the thing at the bottom of Pandora's box was hope. When she's released all of the pain and agony into the world, she holds up hope and thinks, this'll make it alright! There were, in fact, two boxes, one full of beneficence, the other full of malevolence. Hope was in the malevolent box. Right at the bottom, where you have to scrape. So, what is this complex emotion, and does it ever do us any good? Some scholars interpret hope as the cruellest of the curses, forever promising good but always delivering torment to the expectant but passive soul. Others believe it to be the greatest of blessings, the energy of the Earth wrapped in a single feeling, an anticipatory positivity that drives us on.

The climate crisis is not a single-issue problem. Halt the rising temperature, but what about biodiversity? Get an electric car, but what about the lithium mines? Save Alaskan king salmon, but what happens to the fishing communities? One group of thinkers cannot solve all of this alone. Scientists, politicians, conservationists, activists, artists, humanitarians, citizens... all of us have our part to play in working towards solving these interconnected problems. The single thing that unites us all is hope.

Could there be some kind of hope that is dangerous? Yes. When aliens land on our lush and recovering planet in several thousand years they will find evidence of a people who died waiting for the help that never came. We have to stop secretly hoping that a grown-up will step up and fix this thing. Maybe we need to admit that Pandora's box is the pit of our stomachs where all nascent fears begin before they take wing, and put hope there, firmly and indefatigably back where it belongs. Hope in ourselves is our only hope.

In this chapter we celebrate hope in its many forms, from the exuberant energy of Kevin Dyer's *Don't Worry, Be Happy*, about a group of young environmentalists putting on a play about the climate, to the quiet intelligence of Efua Sutherland's *Foriwa,* offering a window into how community diplomacy can gently move everyone forward.

92 THE MAD WOMAN OF CHAILLOT
Jean Giraudoux (1945)

'Those who impoverish the earth, who steal our boas, who prepare for war, who take commissions, who get themselves named to posts without being qualified, who corrupt young people, are going to be here, together, in this room. Do we have the right to eliminate them in one fell swoop?'

This is an overlooked satirical gem, a prescient climate play powered by the determination and guile of a ridiculed older woman. It has everything: unexpected heroes, evil industrialists, a utopian vision, love, harmony, wit and Parisian café culture. You can almost smell the Gauloises.

Two financiers meet for coffee and discuss ways to short the market. (Giraudoux wrote this play during the Second World War when profiteering out of other people's misery was rife.) The waitress, the rich variety of street characters and the local policeman are belittled by these masters of the universe. The Broker and the President are joined by the Prospector and the Baron to discuss their schemes. They are overheard discussing how to exploit the mineral wealth of France. Word of this gets to Countess Aurelia, the unconventional grand-dame who lives in the neighbourhood. She immediately recognises this exploitation for the greed and vandalism it is.

Aurelia is a romantic who has always believed happiness will prevail. She cares for the poets and the animals and all of the rich variety of Parisian life. However, after conversations with her more worldly friend the Ragpicker, she realises she must alter the course of events. She becomes determined to stop the Prospector, the President, the Broker and the Baron before they can begin. She contrives to get them all together in one room and despatch the lot. She enlists the help of the marginalised crew of outsiders and idealists who populate the city. She holds a surrealist tea party with her female friends in which they do a mock-trial of the corrupt businessmen. Aurelia sends word to the men that she has discovered oil under the streets of Paris. One by one they come, rubbing their hands in glee. Each is cast into the sewer, never to be seen again. Instantly the grey urban grime lifts from the city, the birds sing and fly, the plane trees start podding, strangers hug each other, a new flourishing utopia is born.

Aurelia, the eponymous mad woman of Chaillot, is deemed off-centre, or eccentric, by those who disregard kindness and community. Her worldview is seen as out of order by those who have reordered the world to make it easier to exploit. And most effectively she is labelled 'mad' by the misogynists who silence older women and create tropes like 'cat lady'. But she has remained rooted in her community and in herself, as the capitalist world shifted around her, decentring her, and distorting how she is perceived. This simple act of believing in her neighbours and her neighbourhood turns her into a prophet. She can sense that too much power is being given to those who destroy and not enough given to those who protect and create.

The capitalists in the play have subjugated everyone they meet, they have taken from the poor, they have even taken a cut out of relief packages meant for those affected by environmental crises in other parts of France. Enough is enough. Aurelia is sick of their eco-hubris, they are the people who cause the damage and then believe they are above the consequences.

The Mad Woman of Chaillot is bursting with the physical experience of living in the city, the sights and smells, the noise and the dirt. It is a play that feels thrillingly alive. What would cause anyone to destroy this? The ecosystem of Aurelia's neighbourhood, the people, the life, the street-dwellers and flower-sellers, the birds, trees and cats. This is her understanding of community, a place of thriving, mutually beneficial, contented coexistence.

She'll be damned if they take it away from her.

93 MY OCEAN
Sasha Singer-Wilson (2015)

'I think about the ocean, the deep, intelligent ocean that's inside me all of the time and I know that everything is going to be okay.'

Aided by a karaoke machine and a jellyfish dance, Lenny is the hero of Sasha Singer-Wilson's perfectly pitched monologue.

My Ocean is the title of twelve-year-old Lenny's presentation about the decline in the sea turtle population. He has entered a Speaker's League competition and this is his pitch. He starts off confidently enough, relying on facts and figures about the sea and its poorly turtles, but then gets sidetracked. A different truth breaks through. Lenny reveals that he is estranged from his friends and family, and his story is heartbreaking. He has found companionship in the sea rather than in people, and although he embodies sweetness and charm, there is a fire burning deep inside Lenny. He tells the story of mankind's destructive impact on the oceans, and in doing so he also reveals the truth of his own violent and destructive actions. His courage to acknowledge his demons and step up to face the consequences is a rallying call for us all to do the same.

Sasha Singer-Wilson's monologue is a very funny and moving story of one boy's attempt to make sense of the world. Beautifully targeted, it places centre-stage a child's narrative of what it is like to grow up in the twenty-first century. The audience are asked to put themselves in Lenny's shoes, to understand his past and see the future he is growing up imagining.

Distilling the vastness of climate anxiety into a single child's voice proves to be an extraordinarily powerful dramatic coup. *My Ocean* is a micro-fable, a beautiful, warm-hearted piece for a soaring solo voice.

94

~~NOT~~ THE END OF THE WORLD OR ~~KEIN~~ WELTUNTERGANG

Chris Bush (2021)

'Well, of all the reasons for leaving an employer... I have to say, "my boss was eaten by bears" is a new one on me. So you're looking for a fresh start?'

This audacious play takes many of the individual issues that make up the climate crisis and nimbly works them into a multiverse, allowing us to swim in and out of our indecision, evasion, beliefs, intentions, guilt, optional futures and possible pasts.

The concept of the multiverse is both a scientific line of enquiry and a philosophical debate. It imagines alternative realities with different futures, infinite versions of ourselves and endless possibilities. Chris Bush's play explores how we are robbing ourself of a future by having such a hyper-stimulating present. Might the quantum possibilities in our imaginations actually be inhibitors of change?

There is also real human feeling in the play as four characters strive to understand their place in the world, both past and present. Anna is being interviewed by leading climate academic Uta for a position on the professor's research team. Lena is delivering a eulogy for her mother. Years later, Lilly is interviewing Anna about a woman found dead on the ice. All these threads are woven together in a series of pithy and explosive encounters. Written in 228 short scenes, some only one line long, the women's stories are intercut throughout, moving backwards and forwards through time and across alternate realities. There are repetitions with subtle changes and repetitions with wholesale changes. The play moves in and out of the micro-detail, which is forensically rendered, as well as in and out of the meta-philosophy, expansively and generously shared. This is writing that has you sensing the cold air atop the ice pack, and astral plane-ing towards the outer envelope of your imaginative capabilities.

~~Not~~ the End of the World or ~~Kein~~ Weltuntergang is well versed in both climate science and its accompanying philosophical debates, and explores how the richness and complexity of such thinking can be both liberating and paralysing. Wild philosophical questions land out of left field in easily digestible and relatable form. It is elegant in its effortless reach towards a vast scale of thinking whilst remaining human, personable and grounded. It speeds up our collective quest to have the big conversations now, and to know ourselves more thoroughly in the process. The theatricality of the multiverse is used to full and potent effect. There is still time for the human race to fix the climate crisis. Or is there? This is an enriching and enriched piece of theatre.

95 DON'T WORRY, BE HAPPY
Kevin Dyer (2020)

'The old uns are a lost cause. They've had their chance to save the planet and they've wasted it.'

A cast of environmentalists take centre stage in this smart, theatrical play from Kevin Dyer. Taking it in turns to pedal a bike to power the action, the young characters welcome the audience to their carbon-neutral show. They promise us a cracking story, and don't worry, they won't make us feel guilty. They hint that they may give us a soppy polar bear song at the end. After all, that's what audiences want from climate-change theatre, right? *Don't Worry, Be Happy* is a kind, inventive, witty play about how to be happy whilst saving the world.

The action of the play happens in three different times and places, pivoting around an unidentified catastrophe.

Before it happened: The young environmentalists decide to take direct action. They each agree to *do* something – chain themselves to a tanker, pour sugar in someone's brand-new BMW, turn off the freezers in a supermarket – and face the consequences.

When it is happening: Salt water is rising up from under the ground on a Pacific island, the water table is compromised, there is very little time left. A boy tries to build a wall out of Lego to hold back the impending waves, the ocean animates itself in his imagination as a monster.

After it has happened: A group of workers pick up stones in advance of the Planting, an attempt to grow food once more on dilapidated soil. They have to be careful. They could get shot if they pick up something they shouldn't. They have heard rumours of a previous world here, one which had great cities now lost underwater or landfill. Could that be true? They speak in fractured language and with limited vocabulary.

This play trades in a wealth of fascinating ideas about time and language. Eco-linguistics is a field of study that explores the links between the words we speak and the living world we inhabit. Is human language shaped by the natural world it arose from? Language evolved at a time of abundance in nature and complex societies developed intricate words to describe specific natural phenomena. In return, the basis on which human beings engage with the natural world is predicated on the language available to describe it. Some cultures do indeed have fifty words for snow, or rain, or shade, because that specificity of vocabulary reflects a quality of noticing, understanding and careful action. Most societies no longer know the breadth of vocabulary they used to have for nature. Words fall out of use. Does language itself have the power to inspire us to protect, or to neglect our environments?

The decline in biodiversity is linked to the decline in world languages. Areas of the globe that are rich in language, such as Papua New Guinea, are also rich in biodiversity. As habitats in these areas are degraded and cleared, so language too is lost. The UN calculates that up to 90% of world languages could be lost by the end of this century. The paucity of vocabulary recreated in the play is pertinent and chilling, reflecting as it does the scarcity of nature.

Whilst the play is being performed, the cast are still living their normal teenage lives. The play-within-a-play structure allows us to glimpse a love story, exam pressures, anxiety and someone's parents splitting up. It makes abundantly clear the responsibility we elders have to support these young people to live their fullest lives.

And there is, as promised, a soppy polar bear song at the end. But it comes with a warning: '*If you have been affected by any of the issues in tonight's play… do something about it, eh?*'

96 THE CAUCASIAN CHALK CIRCLE Bertolt Brecht (1948)

'The valley to the waterers, that it shall bear fruit.'

Bertolt Brecht's timeless parable celebrates the moral triumph of stewardship over ownership. The Earth and all its bounty should belong to those who will use it most wisely, not those with the fattest wallets.

The Prologue tells the audience everything they need to know about how to watch the play. Two communities are vying for the right to farm a valley. The fruit growers believe it should be their land; the goat herders stake their rival claim. After hearing each other out, both sides agree the fruit growers have the superior claim; their plans to irrigate the valley to provide long-term sustenance for both communities are thoughtful and detailed. To celebrate their agreement, the two sides decide a story should be told. They settle down to watch and learn from the retelling of an ancient Chinese play, *The Caucasian Chalk Circle.*

A grossly corrupt governor is overthrown in a military coup. As he and his wife Natasha flee their city, they leave behind their only child, Michael. He ends up in the hands of a peasant girl, Grusha, who feels she has no choice but to leave her soldier boyfriend Simon and carry Michael to safety. Hunted down by soldiers, she undertakes an epic journey into the mountains, overcoming physical and psychological obstacles along the way, including embarking on a misguided relationship in an attempt to safeguard the child's future. Betrayed to the Ironshirts, she takes Michael and runs. In order to reach her brother she must cross a ravine on a rickety bridge high up in the glacier. There she finds judgement rather than welcome and is forced to marry an elderly man as protection for both herself and Michael. Many months later Simon tracks her down. Both heartbroken, it is impossible for them to share with each other what they have been through. The fighting is over, the governor is restored and Grusha must now return home. A trial is held, presided over by the anarchic, amoral judge Azdak. He suggests a tug-of-war with Michael's birth mother, the corrupt Governor's wife, to prove who is the real mother. A chalk circle is drawn on the ground and the child is placed at the centre. The women are invited to take an arm each. Whoever pulls him to them wins. Grusha refuses, and by doing so reveals herself to have the child's best interests at heart. She is awarded custody. And gets Simon back too.

Every scene in Brecht's play connects to the natural world. From the river that separates Simon and Grusha in the most heartbreaking love scene, to the arduous foraging Grusha must do to feed Michael and find him milk, place and labour are the twin heartbeats of the play. This is earthy, mud-under-the-fingernails kind of theatre. Brecht proves himself an astute environmentalist, addressing the beauty and reality of the natural world with eloquence and care. The working poor are shown to have a deep understanding and respect for the environment, and an empathy for their fellow human beings. He contrasts this with the landed elites and their illusion of nobility. The simple moral that the land, and the child, shall go to those who will care for them most wisely, is perhaps the only guide to living we need. Ownership is nothing, custodianship is all.

97 HEAVY WEATHER
Lizzie Nunnery (2020)

'You're scared of kids who know how bad things are and just how much you've messed things up.'

This is a rite-of-passage play for a young audience. Through ensemble theatre, song and chorus, it speaks in the voice of the Vanessa Nakate generation and asks their most painful questions. How does a child live through the moment when they become disillusioned with their parent? Mona sees her mother making the loudest fuss, acting out the life of an activist, claiming to be 'doing something' about the climate emergency. But she soon learns her mother is all image and no substance. The realisation is an arrow to Mona's heart. But what emerges is a brave and true young character who lives through this familial fire, and steps forward into genuine activism with real weight.

After a panic attack in a science lesson, Mona is shocked into paralysis. The subject of the day is climate science and the impending disaster. She stays in her room, refusing to eat or drink, doom-scrolling through social media. Her timeline is a mix of classmates' cruel gossip and pessimistic climate-change videos. An ensemble cast perform these internet trolls and influencers, constantly chatting into Mona's ear. Classmates laugh at her. Green influencers seem less concerned about the planet than their number of followers. Mona is in a tailspin with all the contradictory and confusing information.

Greenwashing is when businesses or organisations claim to have impeccable eco-credentials whilst in reality continuing with damaging practices. Clamping down on this practice is a growing part of the discourse about the climate emergency. Billions of pounds are spent by

PR companies to obfuscate a company's message, designing logos that just happen to be green, promising integrity, rebranding consumerism to ease guilt but in reality changing little. The vocabulary of 'offsetting', 'carbon neutral' and 'net zero' has only just been invented, but is already vulnerable to disingenuous use. Carbon-neutral petrol, anyone? Governments and advertising watchdogs are frantically scrambling to set up legal standards to hold those who make unsubstantiated or exaggerated claims about the eco-ness of a product to account. It's no surprise that Mona finds herself paranoid and anxious.

Mona is looked after by her big sister Elin because their mum left the family home years ago to run around the world with climate activists. One day Mona sees her mum in the background of a photo of a protest and decides to find her. The outside world is as scary and as exaggerated as the online world. When Mona finally finds her mum, she realises she is a phoney. Her mother leaves her behind again to go and party at the Paris protests. Mona has a breakdown, observed, recorded and commented on by the ever-present ensemble, the eyes of the internet.

But Mona has a realisation. Protesting is not about being seen to protest. Mona fights back, raw and honest, and what goes viral is a young woman emerging as a symbol of hope. She and Elin determine to create their own future.

Heavy Weather is written with great humour, a lightness of touch and emotional integrity. Mona is a relatable and engaging guide through the online climate-change maelstrom. The play has original songs throughout, with the final song – '*I'm waking up, I'm breaking free / Hear me, won't you hear me?*' – being a great moment of uncompromised optimism.

EASTER 98
August Strindberg (1901)

'Oh, I feel already it has cleared for beautiful weather; that the snow is melting. Tomorrow the violets will bloom by the south wall. The clouds have lifted – I feel it in my breathing.'

This beautiful play from one of Europe's finest dramatists is about redemption and the emotional renewal of spring. It takes a set of discordant relationships under intolerable pressure and retunes them, allowing them to find genuine harmony.

The Heyst family are in a bad way. Mr Heyst has been imprisoned for fraud, leaving his family destitute. His wife is in denial about his guilt. Their teacher son Ellis is depressed, lacking professional and social confidence. Their daughter Eleanora has also been incarcerated, but in an asylum, suffering from the mental distress of her father's disgrace. Ellis and his mother share their home with Benjamin, an orphaned student whose inheritance Heyst was responsible for losing. The family fear the debt collector's knock on the door.

Eleanora escapes from the asylum and comes home, incidentally stealing an Easter lily from a florist on the way. The police have been called. So far, so Strindberg, so miserable. But, over the weekend, from Good Friday to Easter Monday, the literal and symbolic clouds begin to lift from this beleaguered household. The snow outside melts and the sea mist clears. Benjamin is entranced by the radiant innocence of Eleanora. The warmth she brings proves irresistible to everyone. She is a catalyst for change, her lightness coinciding with the emergence of spring. Her mother's denial gives way to understanding, and her brother's defences begin to thaw. Lindqvist, the debt collector, searches within himself, and finds the grace to offer redemption.

Eco-consciousness is a state of deep environmental awareness. It is the move beyond latent awareness, into perception. Strindberg nurtures his characters into this condition. The pagan/religious rituals of Easter unfold around them, in tune with the eternal rhythm of the seasons. The coinciding of these rites, the intuitive layering of rituals on top of seasonal change, and the deep connections between the environment and the human, power the deep subtext of his play. Rabindranath Tagore's *The Cycle of Spring* explores similar territory. The characters' eyes lift up to the horizon, outside of themselves, finding healing and forgiveness for, in and of each other, daring to believe in a second chance. *Easter* has profound tenderness, a loving play full of mercy, privileging the enormous importance of being at one with the world, believing that what is to come will be good.

FORIWA 99
Efua Sutherland (1962)

'Is this the way to praise our forefathers? Watching their walls crumble around us? Letting weeds choke the paths they made? Unwilling to open new paths ourselves, because it demands of us thought, and goodwill, and action?'

This is a deceptively simple play about how to gather goodwill and consent for neighbourhood change. *Foriwa* is a classic story ripe for a progressive, green revival.

Efua Sutherland was a pioneer of Ghanaian theatre. A teacher, writer, campaigner, poet and children's advocate, she shaped education, culture and policy. Parks, halls of residence, streets and literary prizes are named after her. Collaboration was her modus operandi, her natural default, embracing pan-Africanism to encourage positive change across cultural and physical borders, including reaching out to the African Diaspora around the world. This spirit of inclusiveness and generosity is present in *Foriwa*.

The titular central character Foriwa is from a town called Kyerefaso, where her mother is Queen Mother, the town leader. Queen Mother is progressive and excited about change but frequently butts heads with the town's elders who are stuck in the past, clinging to arcane traditions, refusing to notice that the town and its citizens are both lagging behind and falling apart. Home from university, Foriwa befriends Labaran, a young stranger from the north. He lives on the streets and isn't from around here, yet he tries to encourage the citizens to clear weeds, sweep up rubbish and take pride in their town. As preparation for the town's festival approaches, a clash of ideologies, and generations, is inevitable. But Queen Mother navigates this split with adept diplomacy, listening deeply and responding thoughtfully. At moments she is direct and critical, at others galvanising and inspirational. She is always patient, reassuring and kind. Hers is an exemplary route to consensual change.

The simple act of encouraging a community to clear up its rubbish is the first step to asking it to care for the environment and for each other. Self-care leads to community-care. This is replicated in communities around the world. Litter-picking and beach-cleaning, simple acts of local activism, become a deeply political act for people who feel ever-more connected. It becomes the progressive act of civic connection they were missing. *Foriwa* brings old traditions into alliance with new thinking, encouraging cooperation and goodwill. It dramatises a female-led social engagement, that allows swift and decisive positive change to evolve. Establishing a community bookshop to educate the town's young, and the setting up of investment in a cooperative plantation brings much-needed pride, identity and income to the town's workers. Hope is the simplicity of that first step.

100 HOW TO SAVE THE WORLD WHEN YOU'RE A YOUNG CARER AND BROKE Nessah Muthy (2021)

'Once I've helped Mum into the bath and bed I'm not getting to bed myself until like midnight and I'm late for school, late with my homework and teachers are getting at me and then I have to rush back to get the slip from the GP because the benefits STILL haven't come through, then I have to leg it to the food bank, just before they shut but I'm landed with another vegan food parcel… and then it gets me thinking, it's a fucking privilege to be a climate activist.'

Of course, Lavisha cares about the planet but she's also got her mum to look after, her schoolwork to do, and a tight household budget to manage. How the hell can she find time to protest and plant fucking trees when she's got so much going on? Nessah Muthy's brilliant and ballsy play takes on class and the climate crisis with energy, wit and an inspiring dauntlessness.

The climate crisis is a class issue. It is the poorest people in the poorest communities who are exposed most to environmental damage and who feel its effects most keenly. Climate activism is also a class issue. Wealthy people can afford to buy heat pumps and go vegan, not to mention take days off work or school to go on protests, secure in the knowledge that neither their education nor their employment will be adversely affected. Not everyone is so privileged. *How to Save the World When You're a Young Carer and Broke* follows Lavisha's personal journey as she wrestles with what she can and can't do to help her mum, the world, and herself.

Forced to move house due to benefits cuts, Lavisha finds herself in a new school where she meets Avril, a posh and proactive fellow student: '*This is the lunch hall! Hashtag humble brag, you see those wooden forks over there? I got them implemented.*' Caught between admiring Avril and finding her sense of entitlement unbearable, Lavisha struggles to admit the truth of her circumstances to her new friend. Avril's admiration for Lavisha's charity-shop clothing doesn't recognise that it is out of economic necessity rather than cool activism. Lavisha's mum suffers from lupus, a debilitating condition causing extreme pain. With this story and Lavisha's care and concern for her mum, Nessah Muthy highlights the vulnerability of people with disabilities in the face of extreme climate conditions. She reminds us that human health is one of the first casualties when the very air we breathe starts to break down. Heatstroke, dehydration and air pollution all disproportionately impact those with weakened immune systems. This is an aspect of the climate crisis not often talked about.

After initial attempts to recruit Lavisha to her various world-saving initiatives, Avril begins to understand that Lavisha's life is very different to hers. They bump into each other at the food bank one day, Avril volunteering and Lavisha collecting. A vegan food parcel might not be exactly what Lavisha needs right now, but it allows them to share a deeper understanding of each other's worldview. Avril opens up about her own childhood as a climate migrant, fleeing cyclones in Bangladesh. Together, the friends decide to take bigger and bolder direct action. They plan a protest at parliament, '*cycloning*' red paint all over the Home Secretary. Realising she cannot risk being arrested and leave her mum to fend for herself, Lavisha finds herself torn between personal responsibility and direct activism.

Lavisha is a charismatic and joyful narrator, full of relatable teenage humour, anger and anxiety. She's also an extraordinary force of nature, a young woman on a mission to discover a cure for lupus as well as do what she can to make the world a better place. This is a loud, funny, exuberant play packed with catchy songs and strong emotions. The audience are invited to contribute props and lighting effects as needed, a sustainable no-waste production that joyfully includes everyone in its making. *How to Save the World When You're a Young Carer and Broke* ends with a powerful call to arms for the audience (whose energy has collectively helped create the show) to carry on with this community-building, to experience the adrenalin rush of direct activism outside the safety of this theatre, to do whatever they can – however big or small – to make their kind of difference. Perhaps our individual actions can add up to a collective whole, a future-altering wave of change, formed of powerful individual drops.

ADDITIONAL PLAYS

101 HEROES AND SAINTS
Cherríe Moraga (1994)

'They throw some dirt over a dump, put some casas de cartón on top of it y dicen que it's the "American Dream." Pues, this dream has turned to pesadilla.'

Cherríe Moraga's influential play about environmental degradation and social oppression continues to pierce and startle today. An ecofeminist parable inspired by real-life events that took place in Delano, California, the play is both accurate reportage of circumstances endured by the Mexican American community in that city as well as an allegory for colonial exploitation and the ongoing and pernicious legacy of its human and environmental costs.

The United Farm Workers' strike was a demand for political and social freedom by the mostly Chicana/o workers. The protest was brought to national attention by the press coverage of hunger striker Cesar Chavez, and it is this struggle which focuses and inspires the events of the play. The story is rooted in the unreported consequences of industrial chemical use. Grape workers were poisoned by inadvertent exposure to agricultural pesticides. This wreaked havoc amongst their communities, leading to a high rate of miscarriages, an increase in children born with life-altering conditions, and a significant cancer cluster amongst the adult population. We watch in horror as the play's characters experience the futility of trying to stay healthy by drinking bottled water and taking cold showers. (This play can be read as a fascinating companion piece to *cullud wattah*, the play about the Flint water crisis featured on page 117.)

Amongst a strong cast of predominantly female central characters, Cerezita Valle is the standout theatrical creation. She is a head born without a body: a real character in the narrative of the play, representing the hundreds of children born with limb and body differences, but also a seer of immense psychological and emotional power and presence. It is she who inspires the farm workers to further resist exploitation. It is she who directs the children to perform an extraordinary act of creative activism – to hang the bodies of those who have died on cruciform scaffolds across the vineyards. These become a series of real-life icons painted on what turns into a vast geopolitical canvas.

This use of sacrificial imagery, drawing on both Christian and ancient South American cultures, infuses the play with a deep mythic power. Cerezita is at the very heart of this alchemy, including a transgressive moment where she embarks on an emotional relationship with the local Catholic priest, Father Juan. This reaching across worlds is a courageous and symbolic act, and it is matched elsewhere in the writing by Moraga's interrogation of the many layers of oppression: internalised female suppression, Catholic homophobia, and the faceless corporations who are the root cause of the environmental and social damage experienced by this community. Women are at the centre of this play, their bodies, their babies and their power. Amparo is the hero of the title, who we watch grow into a formidable activist.

The playwright is herself a Chicana activist, artist and educator. This two-act play uses a bilingual and colloquial mix of Spanish and English, with characters switching languages spontaneously and fluidly. This linguistic heartbeat gives the play a distinct poetry and rhythm. Feelings of anger, pain and fear are beautifully expressed in both languages. Moraga also includes, in the tradition of Indigenous oral storytelling, the collective character of el Pueblo – 'the people, the children, and the community' – which vocalises their growing political awareness. As the emotional pitch of the play rises and the community's anger towards both the Church and the agricultural landowners increases, the play breaks form and moves thrillingly between personal character revelation, magic realism and civic activism. Crop dusters continually fly overhead, until, in a scene reminiscent of Hitchcock's *North by Northwest*, shots are fired and Juan and Cerezita are killed. The people, enraged by this attack, set the fields alight and a huge fire engulfs the stage.

Notable for its confidence to inhabit both realism and surrealism simultaneously, against a backdrop of fierce political activism, *Heroes and Saints* is a play with passion and heft. It has already deservedly earned its longevity in the playwriting canon, and it demands to be performed into the future widely and repeatedly.

102 THE INVENTION OF SEEDS
Annalisa Dias (2022)

'They changed the DNA, so I guess that's how they claim it's an invention.'

Can you copyright a seed? That's the billion-dollar (literally) question at the heart of this intriguing play.

Jessie, a non-binary teenager, lives with their father David on the family farm in Indiana. David is a fourth-generation farmer, and he eats what he sows. He is proud to grow soybeans, corn, wheat, tomatoes and cover crops like alfalfa and ryegrass. Jessie dreams of escaping to art school; their college fund is loaded up and it's time to put together a portfolio for applications. Enter a representative of iGrow, a huge agri-tech company, accusing David of patent infringement. He has used some of the company's seeds without licence. Jessie finds the college fund swiftly swallowed up in lawyers' fees as David (again, quite literally) takes on Goliath. All Jessie can do is make art to bear witness to the Kafkaesque nightmare the family now find themselves in.

The context of the play is the legislation that permits the patenting and copyrighting of manipulated genome sequences in organic matter. North America now holds the largest share of the global agri-tech market and is anticipated to continue that dominance into the coming decade. Biochemicals, next-generation DNA sequencing, robotics, sensor-based technologies, genome editing, bio-energies, blockchain trading, digital monitoring, not to mention biofuels, yield-improvers, resilience-deepeners... With an increasing global population, unpredictable growing conditions, and an ever-destabilising geopolitical outlook contingent on that climate unpredictability, the race for national food security and increased farming efficiency has very definitely begun.

If a bio-tech company genetically tweaks a seed to make it more tolerant in drought conditions, does that mean they have created new intellectual property worthy of protection? When did the penknife-grafting know-how of farmers and horticulturalists become a tradable part of the free market? Should we be able to patent and license these minute adjustments to the natural world? And just how powerful are the legal protections in this corner of the innovative foodstuffs business? Annalisa Dias's play stakes out a world in which the audience can explore the real-world consequences of this historic step in legislation. Does siphoning control of the global harvest into the hands of big business raise intractable ethical questions? If the world faces an increasingly hungry future, will we find ourselves in direct opposition to a free-market balance sheet that trades big on commodity shortages and relies on volatility?

The Invention of Seeds dances around such questions with poise and clarity. While David is being investigated by iGrow, Jessie undertakes their own research. With a young artist's eyes, Jessie straddles the worlds of science and creativity with their illuminating investigation. Jessie grapples with the idea that at some point, nature's invention became classed as 'technology'. They become outraged that their father is being treated like a criminal, but when it is pointed out to them that iGrow is involved in sequencing soybeans that could help alleviate hunger in the global south, they find the empathy in such a paradox. The character of Dr. Gugu Khumalo, a bio-geneticist at Purdue State University, is sanguine about the morality of her research into such miracle seeds being funded by iGrow – the company's business model depends on patents, and in turn, it is that incentive that fuels research. As Jessie delves deeper into the conceptual ideas that are behind David's predicament, the art of nature and the nature of art become intertwined.

The play suggests puppetry to give voice to the seed at the heart of the legal action, Soybean A498Y34.2035. Like Hallie Flanagan Davies in *E=MC²* (page 74), this writer steps into the epic form of the story by creating characters out of more-than-human beings; the Soil, a Crow and the Centre of the Earth speak to us of the breadth and dimension of these questions. We journey through time and subterranean space with this beguiling cast of characters, all trying to figure out what it means to hold the latent potential to nurture and nourish other lives in your hands. Together, they ask us to wonder about the ethics of monetizing that effort.

MAMMELEPHANT 103
Lanxing Fu (2022)

'I don't mind playing God in the least. We are already doing it, why not do it better?'

A reluctant conservationist finds himself standing on the melting permafrost in Siberia trying to persuade a mammelephant to do its thing. Go and roam the tundra! Stomp the snow! Knock the trees down! Keep the cold locked in, keep the sun's heat out, and please keep the earth's temperature stable! Nikky wants to save the world, or rather, his father wanted to save the world, and now that he's died, the son feels beholden to the father's work. However, this new creature, a genetic mix of modern elephant and extinct mammoth, is having none of it. An eloquent and philosophical being struggling with the deep loneliness that comes with being the first and only of its kind, the mammelephant doesn't feel that it owes humankind anything.

Lanxing Fu's elegant, inspired and inspiring play, first produced for the climate justice theatre group Superhero Clubhouse, presents a compassionate worldview pleading for equitability between species – human, animal or otherwise. After a searching conversation with Nikky about their ostensible responsibilities for the future of humankind, Mammelephant wanders across the conservation park that they are forced to call home. They find a friend in Bighorn Sheep, a fellow inhabitant of this rewilded grassland, and share a memorable scene with Yakutian Horse, discussing the chewiness of particular grasses, much like a discussion of a fine wine. Poor Bighorn Sheep then falls victim to one of the hazards of melting permafrost, swallowed up by the semi-regular landslides the area experiences now that the ground is no longer stable.

Pleistocene Park is a real geoengineering project situated in Sakha, northern Siberia. Named after the geological epoch that ended just twelve thousand years ago (having begun over two million years earlier), it is an attempt to recreate the cold-weather savannah-style grasslands that characterised the time known as the Ice Age. Although this period is better known for its deep chills and thick glaciers, huge swathes of land across the Arctic belt, including Alaska, Siberia and much of the Yukon in Canada, were vast cold-weather plains. This so-called Mammoth Steppe was a prime habitat for megafauna, the large species like mammoths with whom early humans shared the earth.

Written for a cast of four playing multiple roles, actors introduce themselves to the audience and explain how their various characters look and behave. Audience members are invited to join in with dance, music and chanting, all inspired by traditional Sakha song-poems, and drawing authentically on Yakut language and culture. The full-company role-play scene which imagines the circumstances that may have led to the death of the last ever woolly mammoth is both hilarious and moving. Humans are told by animals exactly what their responsibilities are, while animals discuss amongst themselves the impact of knowing the extinction clock is ticking.

Funny, provocative and interactive, *Mammelephant* successfully asks the audience to imagine how it feels to play your part in something bigger than humanity, and to dream of something larger than the experience of your own generational timeline. It dares us to imagine our species in the context of the earth's history, alongside the history of all the other species with whom we are privileged to share our planetary home. It achieves this mind-expanding trick with an elegant exploration of a precious thought.

De-extinction is no longer just an idea from a film franchise about dinosaurs. Extracting DNA from long-vanished species is now not only theoretically possible, it has been done. We have moved from the realm of hypothesis into the realm of everyday laboratory practice. The last mammoth only died four thousand years ago. Europe was in the Bronze Age, the Egyptians were admiring the pyramids they'd already built; four-wheeled carts were a thing; there were cities in the Middle East; writing was about to be developed; pottery, rock painting, maize production, you name it, humans around the world were already doing it. When archaeologists dig up mammoth bones, they do so mostly in the colder parts of the globe, and the intact DNA is well-preserved and accessible. Woolly mammoths share over 99% of their DNA with the modern elephant. Advances in genome editing are promising the possibility that the woolly mammoth (or something very like it) will be brought back from extinction within our lifetimes. Cue the gold rush as laboratories around the world race to be first to introduce the 'mammelephant' to the world. In the background of this play, the 'why' lingers. The theory is that a large cold-resistant herbivore stomping around the Arctic Belt will decelerate the melting of the permafrost and keep millions of tonnes of carbon deep in the ground. Is this the right justification to, as Nikky says, play God?

As Nikky and Mammelephant muse together on how much this reintroduction project can realistically achieve, the chilling reality of modern science is brought into vivid theatrical life. The play persuades us to look squarely in the eye at the eradication of cultures as well as species, at the displacement of peoples and lifeforms, and at the resulting loneliness caused by the loss and fracturing of ecologies and communities. *Mammelephant* is a majestic and haunting play for our times. Its final image, of the title character setting off on their own journey, entering a low-impact nomadic lifestyle accompanied by a revolving cast of fellow species (including humans), is a lingering and deeply affecting image of a quiet, collective mutuality, caring for each other by thinking beyond ourselves.

104 SALMON IS EVERYTHING Theresa J. May and the Klamath Theatre Project (2014)

'I get sick of her trying to "advocate" for us, telling me how to protect what's already mine!'

This play follows the response of three families in the aftermath of a devastating fish kill in Oregon's Klamath River in 2002, when thirty thousand salmon were found dead or rotting along its banks. What emerges from the play is a deep and abiding sense of kinship between the people, the salmon and the land. The play's relevance today is as stark as it was two decades ago, not only for the communities that live alongside all the great rivers in the world but for anyone who has ever tasted sustenance from our great water sources.

For First Peoples in the Pacific Northwest, salmon really is everything. The seasonal investment in the rhythms of the annual Chinook and Coho salmon's migration is the centrifugal force around which family, food, ritual, culture and identity all revolve. The events that led to the once-abundant Klamath Coho salmon's collapse, and to it becoming listed as threatened under the Endangered Species Act, is a depressingly familiar tale. Environmental mismanagement and a failure by lawmakers to engage Indigenous peoples in meaningful dialogue about their deep expertise and generational knowledge delayed any potential mitigation of the disaster. Instead, a hundred years' worth of corporate disdain for those native to the land, including wilful ignorance of Traditional Ecological Knowledge, resulted in the mass diversion of water for agriculture. This led directly to the entirely preventable Lower Klamath fish kill. (That the phrase 'fish kill' is so familiar is a sobering marker in itself.) Investigations soon pointed the finger at high water temperatures, low water levels and toxic algae, all directly caused by the overuse of water by agricultural industries.

Salmon Is Everything tells this story from multiple Indigenous perspectives, including those of local Hupa people and Karuk, Yurok and Klamath Indian tribes. It was created in consultation and collaboration with those communities, resulting in a culturally wise and sensitive piece of theatre in both its form and content. The story is episodic and multi-vocal, yet it resists fracturing and remains a holistic experience. This cumulation of different perspectives illuminates nuance that would otherwise be lost. Those who speak to us in monologue are composite characters, created from people involved in both the events and the research. It becomes abundantly clear that the century or so of wilful mismanagement is nothing more or less than cultural genocide against these populations. Could corporate genocide ever be on our statute books? The climax of the play is the Longhouse meeting, where all the individual voices come together into a polyphonic statement, a passionate declaration that the rights of those who inhabit the Klamath watershed must have agency in decisions about their shared future.

Both the play itself and the process of creating it confirm the optimism that shared responsibility can lead directly to shared benefit. As the opening quote illustrates, it is also open and honest in its exploration of the role of the non-Indigenous ally to co-create the story.

Salmon Is Everything is a tender and compassionate story, inspiringly so. It finds humour within the heartbreak and captures a bone-deep authenticity about our human relationship to seasonality, migratory populations, communality and the boon of kinship.

SOMEWHERE 105
Marisela Treviño Orta (2020)

'I wish back in October I had known I was eating the last tomato I was ever gonna eat.'

The breakdown of civil society might be not far off, but entomologist Cassandra has only one thing on her mind. Butterflies. It has been her life's work to hatch, nurture and rear what might just be the last monarch butterflies in the world, and even if the apocalypse is on the horizon, she is not going to miss the chance to follow their migration. With her brother Alexander for company, she jumps on her bicycle and follows the butterflies towards the coast, pedalling fast and following a tracker – a tiny antennae attached to the butterflies in the lab – only resting at night when the butterflies stop to catch their breath. The siblings avoid towns (there are rumours of terrible things happening), camp out under the stars, and trust to two things – the butterflies' movement and Cassandra's visions. She's seen the coast in her dreams and believes whatever happens when they get there will be beautiful.

Inspired by the extraordinary phenomenon of the real-life autumn migration of monarch butterflies, this play looks at what happens when patterns of behaviour are no longer predictable. As society begins to fall apart and weather patterns change, the butterflies appear to be following a different route. Cassandra is deeply troubled by the potential significance of this development.

She has a vision of an old farmstead, and sure enough she and Alexander find themselves cycling up to an un-sign-posted, dilapidated farm. They are greeted by the wary inhabitants, a group of people who have left the city to wait out the crisis in self-sufficient survival mode. Diana and Sybil are another pair of siblings, attempting to shore up their survival effort by growing food and measuring dwindling rations. Sybil scours the sky for birds – none seen in a week, which means the secondary consumers in the food web are failing. *'It's happening faster than I thought',* she mutters. Her partner Eph digs for truffles and mushrooms with his friend Corin, but they can sense the supply is running short. Something is wrong. The trees are wilting. The mushrooms are shrivelling. An uneasy relationship of mutual benefit is agreed upon with the strangers – Cassandra's scientific knowledge will help figure out what's wrong with the farm's ecosystem, and she and Alexander will be allowed to share water and food while they wait for the butterflies to roost. We watch relationships fracture as the pressures of eking out a survival existence mount up.

This could all be a little bleak, but there is warmth, wit and invention in the writing, and a gorgeous metaphor for collective dreaming. These characters become a microcosm of society. Orta is inspired by the idea in Greek myth of the seer and the visionary. Each character takes on Cassandra's skill in their own rudimentary way. The sense of the future and the will to make it better become the undeniable song of the play. She offers image after image of the extraordinary world of flora, insects and fungi, infusing the play with a rare beauty. The visual imagery of the sub-soil ecology is richly detailed. The exuberance of the aerial dance of the butterflies takes on a stage presence of its own (it is recommended to have a puppeteer on your creative team). As the humans' own survival instincts become deeper, so too does the dramatic language of the play. The natural world is given its own theatrical agency. A man fuses with a tree. A woman ascends with the butterflies. The play takes on a magical surrealism that perfectly encapsulates this change into a new world.

Somewhere is a beautiful and provocative thought experiment about the end times. Where would you go? What would you need? Who would you be with? Do you plan to wait out the coming catastrophe or attempt to create a new form of existence? Might the very act of planning for a difficult future transform into a Cassandra-like vision of a positive, possible future? If the point of no return might be a bit closer than we thought, perhaps it's worth giving it a try.

APPENDICES

Further Resources

There are many organisations working at the cross-section of culture and the climate emergency. Here are just some of them:

www.climatechangetheatreaction.com, short plays about the climate crisis from around the world

www.culturedeclares.org, an international movement of cultural organisations declaring a state of emergency

www.climatecultures.net, a network of artists and their cultural responses to environmental change

www.writersrebel.com, Extinction Rebellion's writing community

www.juliesbicycle.com, mobilising artists and companies to make practical change to their practice

www.theatregreenbook.com
www.sustainablepractice.org
www.broadwaygreen.com, sustainable theatre-making

www.theatrewithoutborders.com, the intersection of art and environment

www.artistsandclimatechange.com, culture moves the conversation forward

www.artivistnetwork.org, where art meets activism

www.ecostage.online, ecological thinking in creative practice

www.howlround.com, an open platform for progressive theatre activism

www.resilientrevolt.org, activist theatre for climate justice

You could fill your shelves with books that have been written to guide us through this moment. Here are a handful we have been particularly drawn to:

The Great Derangement: Climate Change and the Unthinkable by Amitav Ghosh (Penguin, 2019), about our imaginative failure to understand the severity of the crisis

There is No Planet B: A Handbook for the Make or Break Years by Mike Berners-Lee (Cambridge University Press, 2019), a guide to the impact of everything

Underland: A Deep Time Journey by Robert Macfarlane (Penguin, 2019), perspective-enabling writing about legacy and longevity

Braiding Sweetgrass by Robin Wall Kimmerer (Milkweed Editions, 2015), a guide to Indigenous ecological knowledge and wisdom

The Thunder Mutters: 101 Poems for the Planet edited by Alice Oswald (Faber and Faber, 2006), poetic expressions of care and anger

Slow Violence and the Environmentalism of the Poor by Rob Nixon (Harvard University Press, 2013), the activists of the global south shedding light on the invisible, attritional nature of environmental damage

Silent Spring by Rachel Carson (First Mariner Books, 1962), still a seminal text for the environmental movement

Mourning Nature: Hope at the Heart of Ecological Loss and Grief edited by Ashlee Cunsolo and Karen Landman (McGill-Queen's University Press, 2017), a collection of essays about ecological grief and the process towards positive change

This Changes Everything: Capitalism vs. the Climate by Naomi Klein (Simon & Schuster, 2014), an exposé of political and organisational failings

The Ecological Thought by Timothy Morton (Harvard University Press, 2010), which argues for re-thinking the inter-connectedness of all life

Imagining Extinction: The Cultural Meanings of Endangered Species by Ursula K. Heise (The University of Chicago Press, 2016), exploring the cultural meanings of loss and the role the arts play in defining them

The Anthropocene and the Global Environmental Crisis: Rethinking Modernity in a New Epoch edited by Clive Hamilton, Christophe Bonneuil and François Gemenne (Routledge, 2015), a collection of thinking to help the arts, humanities and social sciences respond to the Anthropocene

All We Can Save: Truth, Courage, and Solutions for the Climate Crisis edited by Ayana Elizabeth Johnson and Katharine K. Wilkinson (One World, 2020), a series of inspirational stories from women on the front line of the climate crisis

No One Is Too Small to Make a Difference by Greta Thunberg (Penguin, 2019), a call to arms from the world's most famous environmental activist

The Good Ancestor by Roman Krznaric (Penguin Random House, 2020), about long-term thinking and legacy

Funny Weather by Olivia Laing (Picador, 2021), which looks at the role art plays in a time of crisis

Islands of Abandonment by Cal Flyn (HarperCollins, 2021), about humanity's resilience

Towards an Ecocritical Theatre: Playing the Anthropocene by Mohebat Ahmadi (Routledge, 2022)

The Uninhabitable Earth: Life After Warming by David Wallace-Wells (Tim Duggan Books, 2019)

Staging the End of the World: Theatre in a Time of Climate Crisis by Brian Kulick (Methuen Drama, 2023)

Parable of the Sower by Octavia Butler (Grand Central Publishing, 2000)

Emergent Strategy: Shaping Change, Changing Worlds by adrienne maree brown (AK Press, 2017)

Saving Us: A Climate Scientist's Case for Hope and Healing in a Divided World by Katharine Hayhoe (One Signal Publishers/Atria, 2021)

And a handful of other resources:

Theatre Communications Group's Climate Action page, *www.tcg.org*, which includes the following:

No Dream Deferred, *www.nodreamdeferrednola.com*

Groundwater Arts, *www.groundwaterarts.com*, and their initiatives:
Green New Theater, *www.groundwaterarts.com/green-new-theatre*
Divest to Invest, *www.groundwaterarts.com/divest-to-invest*

Climate Justice Alliance, *www.climatejusticealliance.org*

Movement Generation's *Just Transition*, *www.movementgeneration.org/justtransition*

Another Gulf Is Possible Collaborative, *www.anothergulf.com*

Broadway Green Alliance, *www.broadwaygreen.com*, and their resources:
Sustainable Solutions for Reopening toolkit, *www.broadwaygreen.com/greener-reopening-toolkit*
Video resource library, *www.broadwaygreen.com/videos*

Julie's Bicycle, *www.juliesbicycle.com*

The Centre for Sustainable Practice in the Arts, *www.sustainablepractice.org*

The National Resources Defense Council, *www.nrdc.org*

The Equitable and Just National Climate Platform, *www.ajustclimate.org*

The Sustainable Production Toolkit, *www.sustainableproductiontoolkit.com*

Climate Change Theatre Action, *www.climatechangetheatreaction.com*

The Arts & Climate Initiative, *www.artsandclimate.org*

HowlRound's Theatre in the Age of Climate Change series, *www.howlround.com/series/theatre-age-climate-change*

American Theatre magazine's 'Theatre and Climate Change' special issue, *www.americantheatre.org/category/special-section/theatre-climate-change*

Superhero Clubhouse, *www.superheroclubhouse.org*

Podcasts: *The Planet Pod, Climate Conversations, Shaping the Future, Planet A: Talks on climate change, The Climate Question, How to Make a Difference, The Emergence Magazine Podcast, How to Save a Planet, Mothers of Invention, Hot Take, A Matter of Degrees, Outrage! Optimism, All My Relations, Green Dreamer,* and *Citizens' Climate Radio*

Films: *An Inconvenient Truth* and its sequel *An Inconvenient Sequel: Truth to Power, Chasing Coral, Before the Flood, Plastic China, Aquarela, To The Ends of the Earth, The Condor & the Eagle, This Changes Everything, Climate Refugees, Can't Stop The Water, Snowpiercer, Beasts of the Southern Wild, First Reformed, Frozen II, Wall-E, Interstellar*

Artists: Olafur Eliasson, John Akomfrah, Agnes Denes, Tomás Saracenco, David Buckland, Cai Guo-Qiang, Isaac Cordal, Naziha Mestaoui, Xavier Cortada, Susanna Bauer, Sean 'Hula' Yoro, Luzinterruptus, Sirintip

Numbers referenced throughout these indexes indicate each play's position within the book, rather than its page number.

Index of Plays by Title

We have included ways to source each play: in some cases, from its originating publisher; in others, suggesting websites that should help you track it down. Out-of-print plays can often be bought online from second-hand booksellers.

We have chosen not to suggest editions or translations of many of the classic foreign-language plays included – and instead encourage you to read around and find your own preferred version.

Publishers' Websites

Aurora Metro Books, *www.aurorametro.com*
Concord Theatricals, *www.concordtheatricals.com*
Currency Press, *www.currency.com.au*
Dramatic Publishing, *www.dramaticpublishing.com*
Dramatists Play Service, *www.dramatists.com*
Faber and Faber, *www.faber.co.uk*
Fair Acre Press, *www.fairacrepress.co.uk*
HarperCollins, *www.harpercollins.com*
Indiana University Press, *www.iupress.org*
Laertes Books, *www.laertesbooks.org*
Mercier Press, *www.mercierpress.ie*
Methuen Drama, an imprint of Bloomsbury Publishing, *www.bloomsbury.com*
New Play Exchange, *www.newplayexchange.org*
Nick Hern Books, *www.nickhernbooks.co.uk*, distributed in the US by TCG Books
Oberon Books, an imprint of Bloomsbury Publishing, *www.bloomsbury.com*
OSU Press, *www.oregonstate.edu*
Oxford University Press, *global.oup.com*
Penguin Classics, *www.penguin.co.uk*
Picador, an imprint of Pan Macmillan, *www.panmacmillan.com*
Playscripts, *www.playscripts.com*
Playwrights Canada Press, *www.playwrightscanada.com*, distributed in the US by TCG Books
Talonbooks, *www.talonbooks.com*
TCG Books, *www.tcg.org*
Vintage Books, an imprint of Knopf Doubleday, *www.knopfdoubleday.com*

Index of Playwrights

Index of Plays by Cast Size

We are advocates for gender-blind casting and so have chosen to group the plays simply by cast size allowing for maximum flexibility of casting.

For a cast of 7

For a cast of 8

For a cast of 9

For a cast of 10

For a cast of 12

For a cast of 15

For a medium ensemble (8–12)

For a large ensemble (12+)

Publisher's Note

(*Originally published in the UK edition, 2021, in an expanded version.*)

As the publisher of a book about how theatre can play a part in tackling the climate emergency, Nick Hern Books acknowledges its own responsibilities in being part of this vital effort – and we're committed to producing every copy of this book in an environmentally sustainable way. We have considered and evaluated our working practices, the printing process and our partnerships to best achieve this goal, whilst acknowledging that, regrettably, manufacturing this book and getting into your hands cannot happen without a carbon footprint to some extent.

Mike Berners-Lee's important and sobering book, *How Bad Are Bananas?: The Carbon Footprint of Everything* (Profile Books, 2010), calculates the carbon footprint of a paperback book as comparatively low: 400g CO_2e for each copy of a book printed on recycled paper where every printed copy is sold. (That's about the same as watching twelve hours of television.) All the same, we have, wherever we can, sought to minimise the impact of this book at every stage of the supply chain.

We've chosen a smaller size than we might ordinarily use for a book of this sort (what's known in publishing jargon as 'B-format'), a binding (paperback as opposed to hardback) and a design (printing in black ink only inside the book) to reduce energy and wastage of materials. We've limited our own office-based printing whilst working on the book, shared edits digitally, recycled all paper (as we always do), and held meetings on Zoom (the Covid-19 pandemic accelerated our practices on this too, of course). These important considerations have already and will continue to extend into all books we publish.

Although we have produced this book in ebook format as well as in print – to ensure it can be read by as many people as possible – we're aware that digital ebooks (or at least the devices required to read them) use significantly more resources and energy than this paperback copy. The essential *Is it really green?: Everyday eco-dilemmas answered* by Georgina Wilson-Powell (Dorling Kindersley, 2021) cites 'studies [which] suggest that you would need to read around 25 ebooks a year for the energy and materials used to produce the e-reader to have less environmental impact than the same number of printed books'.

Overall, our aim is that this book's environmental impact might be considered worthwhile if it inspires productions of the plays it features. We hope that new (and, needless to say, sustainable) productions of these brilliant, foresighted plays can actively and positively contribute to addressing the climate emergency which threatens the planet we are lucky to call our home.

Please read this book.

Please read – and produce – these plays.

And then please pass this book on to everyone you know.

Matt Applewhite
Managing Director & Commissioning Editor
Nick Hern Books

A Note on the US Edition

This book is printed digitally, rather than on a more detrimental lithographic printing press, with an initial print run that meets the number of confirmed orders on hand. All future copies will be fulfilled through print-on-demand technology as orders are received by the distributor. The intention is that every single copy that is printed is sold. Often books that go unsold are destroyed, resulting in wasted paper, wasted energy, added greenhouse emissions, and pulping. This book is printed by Lightning Source, whose printing facility is in the warehouse where books are stored, reducing unnecessary transportation costs. Lightning Source expects each of its paper suppliers to be environmentally responsible as well and does not use paper sourced from endangered old growth forests, forests of exceptional conservation value, or the Amazon basin.

Elizabeth Freestone is a theatre director, creative consultant and environmentalist. She has directed plays for the Royal Shakespeare Company, Manchester Royal Exchange, the Citizens Theatre Glasgow, the Young Vic and Shakespeare at the Tobacco Factory amongst others. She is a former Artistic Director of Pentabus, a new-work touring company. She offers strategic advice and creative and environmental consultancy in both a paid and voluntary capacity for various organisations, as well as teaching and mentoring young artists. She has a Masters degree in Environmental Humanities from Bath Spa University.

Originally trained as a sculptor, **Jeanie O'Hare** is a short-story writer, playwright and project consultant for theatre and film. She has worked for the Royal Court Theatre, the Royal Shakespeare Company, Druid Theatre, and was Chair of Playwriting at Yale School of Drama. Most recently she was the Director of New Work Development at The Public Theater in New York. Her focus is on scouting, developing and producing new writers who tell original and important stories.

Printed in the USA
CPSIA information can be obtained
at www.ICGtesting.com
JSHW080727300923
49337JS00001B/1